Refining the Concept of Scientific Inference When Working with Big Data

Proceedings of a Workshop

Ben A. Wender, *Rapporteur*

Committee on Applied and Theoretical Statistics

Board on Mathematical Sciences and Their Applications

Division on Engineering and Physical Sciences

The National Academies of
SCIENCES · ENGINEERING · MEDICINE

THE NATIONAL ACADEMIES PRESS
Washington, DC
www.nap.edu

THE NATIONAL ACADEMIES PRESS   500 Fifth Street, NW   Washington, DC 20001

This workshop was supported by Contract No. HHSN26300076 with the National Institutes of Health and Grant No. DMS-1351163 from the National Science Foundation. Any opinions, findings, or conclusions expressed in this publication do not necessarily reflect the views of any organization or agency that provided support for the project.

International Standard Book Number-13: 978-0-309-45444-5
International Standard Book Number-10: 0-309-45444-1
Digital Object Identifier: 10.17226/24654

This publication is available in limited quantities from:

Board on Mathematical Sciences and Their Applications
500 Fifth Street NW
Washington, DC 20001
bmsa@nas.edu
http://www.nas.edu/bmsa

Additional copies of this publication are available for sale from the National Academies Press, 500 Fifth Street, NW, Keck 360, Washington, DC 20001; (800) 624-6242 or (202) 334-3313; http://www.nap.edu.

Suggested citation: National Academies of Sciences, Engineering, and Medicine. 2017. *Refining the Concept of Scientific Inference When Working with Big Data: Proceedings of a Workshop*. Washington, DC: The National Academies Press. doi: 10.17226/24654.

The National Academies of
SCIENCES · ENGINEERING · MEDICINE

The **National Academy of Sciences** was established in 1863 by an Act of Congress, signed by President Lincoln, as a private, nongovernmental institution to advise the nation on issues related to science and technology. Members are elected by their peers for outstanding contributions to research. Dr. Marcia McNutt is president.

The **National Academy of Engineering** was established in 1964 under the charter of the National Academy of Sciences to bring the practices of engineering to advising the nation. Members are elected by their peers for extraordinary contributions to engineering. Dr. C. D. Mote, Jr., is president.

The **National Academy of Medicine** (formerly the Institute of Medicine) was established in 1970 under the charter of the National Academy of Sciences to advise the nation on medical and health issues. Members are elected by their peers for distinguished contributions to medicine and health. Dr. Victor J. Dzau is president.

The three Academies work together as the **National Academies of Sciences, Engineering, and Medicine** to provide independent, objective analysis and advice to the nation and conduct other activities to solve complex problems and inform public policy decisions. The National Academies also encourage education and research, recognize outstanding contributions to knowledge, and increase public understanding in matters of science, engineering, and medicine.

Learn more about the National Academies of Sciences, Engineering, and Medicine at **www.national-academies.org.**

The National Academies of
SCIENCES · ENGINEERING · MEDICINE

Reports document the evidence-based consensus of an authoring committee of experts. Reports typically include findings, conclusions, and recommendations based on information gathered by the committee and committee deliberations. Reports are peer reviewed and are approved by the National Academies of Sciences, Engineering, and Medicine.

Proceedings chronicle the presentations and discussions at a workshop, symposium, or other convening event. The statements and opinions contained in proceedings are those of the participants and have not been endorsed by other participants, the planning committee, or the National Academies of Sciences, Engineering, and Medicine.

For information about other products and activities of the National Academies, please visit nationalacademies.org/whatwedo.

PLANNING COMMITTEE ON REFINING THE CONCEPT OF SCIENTIFIC INFERENCE WHEN WORKING WITH BIG DATA

MICHAEL J. DANIELS, University of Texas, Austin, *Co-Chair*
ALFRED O. HERO III, University of Michigan, *Co-Chair*
GENEVERA ALLEN, Rice University and Baylor College of Medicine
CONSTANTINE GATSONIS, Brown University
GEOFFREY GINSBURG, Duke University
MICHAEL I. JORDAN, NAS[1]/NAE,[2] University of California, Berkeley
ROBERT E. KASS, Carnegie Mellon University
MICHAEL KOSOROK, University of North Carolina, Chapel Hill
RODERICK J.A. LITTLE, NAM,[3] University of Michigan
JEFFREY S. MORRIS, MD Anderson Cancer Center
RONITT RUBINFELD, Massachusetts Institute of Technology

Staff

MICHELLE K. SCHWALBE, Board Director
BEN A. WENDER, Associate Program Officer
LINDA CASOLA, Staff Editor
RODNEY N. HOWARD, Administrative Assistant
ELIZABETH EULLER, Senior Program Assistant

---

[1] National Academy of Sciences.
[2] National Academy of Engineering.
[3] National Academy of Medicine.

Acknowledgment of Reviewers

This proceedings has been reviewed in draft form by individuals chosen for their diverse perspectives and technical expertise. The purpose of this independent review is to provide candid and critical comments that will assist the institution in making its published proceedings as sound as possible and to ensure that the proceedings meets institutional standards for objectivity, evidence, and responsiveness to the study charge. The review comments and draft manuscript remain confidential to protect the integrity of the deliberative process. We wish to thank the following individuals for their review of this proceedings:

Joseph Hogan, Brown University,
Iain Johnstone, NAS, Stanford University,
Xihong Lin, Harvard University, and
Hal Stern, University of California, Irvine.

Although the reviewers listed above have provided many constructive comments and suggestions, they were not asked to endorse the views presented at the workshop, nor did they see the final draft of the workshop proceedings before its release. The review of this workshop proceedings was overseen by Sallie Keller, Social Decision Analytics Laboratory, Biocomplexity Institute of Virginia Tech, who was responsible for making certain that an independent examination of this workshop proceedings was carried out in accordance with institutional procedures and that all review comments were carefully considered. Responsibility for the final content of this proceedings rests entirely with the rapporteur and the institution.

Contents

1

Introduction

The concept of utilizing big data to enable scientific discovery has generated tremendous excitement and investment from both private and public sectors over the past decade, and expectations continue to grow (FTC, 2016; NITRD/NCO, 2016). Big data is considered herein as data sets whose heterogeneity, complexity, and size—typically measured in terabytes or petabytes—exceed the capability of traditional approaches to data processing, storage, and analysis. Using big data analytics to identify complex patterns hidden inside volumes of data that have never been combined could accelerate the rate of scientific discovery and lead to the development of beneficial technologies and products. For example, an analysis of big data combined from a patient's electronic health records (EHRs), environmental exposure, activities, and genetic and proteomic information is expected to help guide the development of personalized medicine. However, producing actionable scientific knowledge from such large, complex data sets requires statistical models that produce reliable inferences (NRC, 2013). Without careful consideration of the suitability of both available data and the statistical models applied, analysis of big data may result in misleading correlations and false discoveries, which can potentially undermine confidence in scientific research if the results are not reproducible. Thus, while researchers have made significant progress in developing techniques to analyze big data, the ambitious goal of inference remains a critical challenge.

WORKSHOP OVERVIEW

The Committee on Applied and Theoretical Statistics (CATS) of the National Academies of Sciences, Engineering, and Medicine convened a workshop on June 8-9, 2016, to examine critical challenges and opportunities in performing scientific inference reliably when working with big data. With funding from the National Institutes of Health (NIH) Big Data to Knowledge (BD2K) program and the National Science Foundation (NSF) Division of Mathematical Sciences, CATS established a planning committee (see p. v) to develop the workshop agenda (see Appendix B). The workshop statement of task is shown in Box 1.1. More than 700 people registered to participate in the workshop either in person or online (see Appendix A).

This publication is a factual summary of what occurred at the workshop. The planning committee's role was limited to organizing and convening the workshop. The views contained in this proceedings are those of the individual workshop participants and do not necessarily represent the views of the participants as a whole, the planning committee, or the National Academies of Sciences, Engineering, and Medicine. In addition to the summary provided here, materials related to the workshop can be found on the CATS webpage (http://www.nas.edu/statistics), including speaker presentations and archived webcasts of presentation and discussion sessions.

BOX 1.1
Statement of Task

An ad hoc committee appointed by the National Academies of Sciences, Engineering, and Medicine will plan and organize a workshop to examine challenges in applying scientific inference to big data in biomedical applications. To this end, the workshop will explore four key issues of scientific inference:

- Inference about causal discoveries driven by large observational data,
- Inference about discoveries from data on large networks,
- Inference about discoveries based on integration of diverse data sets, and
- Inference when regularization is used to simplify fitting of high-dimensional models.

In addressing these four issues, the workshop will:

- Bring together statisticians, data scientists, and domain researchers from different biomedical disciplines,
- Identify new methodologic developments that hold significant promise, and
- Highlight potential research program areas for the future.

One or more rapporteurs who are not members of the committee will be designated to prepare a workshop summary.

WORKSHOP THEMES

While the workshop presentations spanned multiple disciplines and active domains of research, several themes emerged across the two days of presentations, including the following: (1) big data holds both great promise and perils, (2) inference requires evaluating uncertainty, (3) statisticians must engage early in experimental design and data collection activities, (4) open research questions can propel both the domain sciences and the field of statistics forward, and (5) opportunities exist to strengthen statistics education at all levels. Although some of these themes are not specific to analyses of big data, the challenges are exacerbated and opportunities greater in the context of large, heterogeneous data sets. These themes, described in greater detail below and expanded upon throughout this proceedings, were identified for this publication by the rapporteur and were not selected by the workshop participants or planning committee. Outside of the identified themes, many other important questions were raised with varying levels of detail as described in the summary of individual speaker presentations.

Big Data Holds Both Great Promise and Perils

Many presenters called attention to the tremendous amount of information available through large, complex data sets and described their potential to lead to new scientific discoveries that improve health care research and practice. Unfortunately, such large data sets often contain messy data with confounding factors and, potentially, unidentified biases. These presenters suggested that all of these factors and others be considered during analysis. Many big data sources—such as EHRs—are not collected with a specific research objective in mind and instead represent what presenter Joseph Hogan referred to as "found data." A number of questions arise when trying to use these data to answer specific research questions, such as whether the data are representative of a well-defined population of interest. These often unasked questions are fundamental to the reliability of any inferences made from these data.

With a proliferation of measurement technologies and large data sets, often the number of variables (p) greatly exceeds the number of samples (n), which makes evaluation of the significance of discoveries both challenging and critically important, explained Michael Daniels. Much of the power of big data comes from combining multiple data sets containing different types of information from diverse individuals that were collected at different times using different equipment or experimental procedures. Daniels explained that this can lead to a host of challenges related to small sample sizes, the presence of batch effects, and other sources of noise that may be unknown to the analyst. For such reasons, uncritical analysis of these data sets can lead to misleading correlations and publication of irreproducible results. Thus, big data analytics offers tremendous opportunities but is simultaneously characterized by numerous potential pitfalls, said Daniels. With such abundant, messy, and complex data, "statistical principles could hardly be more important," concluded Hogan.

Andrew Nobel cautioned that "big data isn't necessarily the right data" for answering a specific question. He alluded to the fundamental importance of defining the question of interest and assessing the suitability of the available data to support inferences about that question. Across the 2-day workshop, there was notable variety in the inferential tasks described; for example, Sebastien Haneuse described a comparative effectiveness study of two antidepressants to draw inferences about differential effects on weight gain, whereas Daniela Witten described the use of inferential tools to aid in scientific discovery. Some presenters remarked that big data may invite analysts to overuse exploratory analyses to define research questions and underemphasize the fundamental issues of data suitability and bias. Understanding bias is particularly important with large, complex data sets such as EHRs, explained Daniels, as analysts may not have control over sample selection among other sources of bias. Alfred Hero explained that when working with large data sets that contain information on many diverse variables, quantifying bias and understanding the conditions necessary for replicability can be particularly challenging. Haneuse encouraged researchers using EHRs to compare available data to those data that would result from the ideal randomized trial as a strategy to define missing data and explore selection bias. More broadly, when analyses of big data are used for scientific discovery, to help form scientific conclusions, or to inform decision making, statistical reasoning and inferential formalism are required.

Inference Requires Evaluating Uncertainty

Many workshop presenters described significant advances made in developing algorithms and methods for analyzing large, complex data sets. However, a recurring topic of discussion was that most work to date stops short of formally assessing the uncertainty associated with the predictions or comparisons made with big data (as mentioned in the presentations by Michael Daniels, Alfred Hero, Genevera Allen, Daniela Witten, Michael Kosorok, and Bin Yu). For example, data mining algorithms that generate network structures representing a snapshot of complex genetic processes are of limited value without some understanding of the reliability of the nodes and edges identified, which in this case correspond to specific genes and potential regulatory relationships, respectively. In an applied setting, Allen and Witten suggested using several estimation techniques on a single data set and similarly using a single estimation technique with random subsamples of the observations. In practice, results that hold up across estimation techniques and across subsamples of the data are more likely to be scientifically useful. While this approach offers a starting place, researchers would prefer the ability to compute a confidence interval or false discovery rate for network features of interest. Assessment and communication of uncertainty are particularly important and challenging for exploratory data analyses, which should be viewed as hypothesis-generating activities with high levels of uncertainty to be addressed through follow-up data collection and confirmatory analyses.

Statisticians Must Engage Early in
Experimental Design and Data Collection Activities

Emery Brown, Xihong Lin, Cosma Shalizi, Alfred Hero, and Robert Kass noted that too often statisticians become involved in scientific research projects only after experiments have been designed and data collected. Inadequate involvement of statisticians in such "upstream" activities can negatively impact "downstream" inference, owing to suboptimal collection of information necessary for reliable inference. Furthermore, these speakers indicated that it is increasingly important for statisticians to become involved early in and throughout the research process so as to consider the potential implications of data preprocessing steps on the inference task. In addition to engaging experimental collaborators early, Lin emphasized the importance of cooperating and building alliances with computer scientists to help develop methods and algorithms that are computationally tractable. Responding to a common mischaracterization of statisticians and their scientific collaborators, several other speakers emphasized that statisticians are scientists too and encouraged more of their colleagues to become experimentalists and disciplinary experts pursuing research in a specific domain as opposed to focusing on statistical methods development in isolation from scientific research. Hero suggested that in order to be viewed as integral contributors to scientific advancements, statisticians could aim to be positive and constructive in interacting with collaborators.

Open Research Questions Can Propel Both the
Domain Sciences and the Field of Statistics Forward

Over the course of the workshop, a number of presenters identified various open research questions with potential to advance the fields of statistics and biomedical sciences, as well as the broader scientific research community. Several presenters illustrated the challenges and opportunities of integrating phenomenological data across multiple temporal or spatial scales. Examples included connecting subcellular descriptions of gene and protein expression with longitudinal EHRs and combining neuroscience technologies and methods spanning the individual neuron scale to whole brain regions. Alfred Hero said that the challenges associated with creating integrative statistical models informed by known biology are substantial because of the inherent complexity of biological processes and because integrative models typically require tracking and relating multiple processes. Andrew Nobel and Xihong Lin discussed the importance of developing scalable and computationally efficient inference procedures designed for cloud environments, including increasingly widespread cloud computing and data storage. Similarly, several speakers suggested that the use of artificial intelligence and automated statistical analysis packages will become prevalent and that significant opportunity exists to improve statistical practices for many disciplines by ensuring appropriate methods are implemented in such emerging tools. Finally, a few presenters encouraged research into methods that could better define the questions that a given data set could potentially answer based on the contained information.

Opportunities Exist to Strengthen Statistics Education at All Levels

Emery Brown, Robert Kass, Bin Yu, Andrew Nobel, and Cosma Shalizi emphasized that there are opportunities to improve statistics education and that increased understanding of statistics broadly across scientific disciplines could help many researchers avoid known pitfalls that may be exacerbated when working with big data. One suggestion was to teach probability and statistical concepts and reasoning in middle and high school through a longitudinal and reinforcing curriculum, which could provide students with time to develop statistical intuition. Another suggestion was to organize undergraduate curricula around fundamental principles rather than introducing students to a series of statistical tests to match with data. Many pitfalls faced in analysis of large, heterogeneous data sets result from inappropriate application of simplifying assumptions that are used in introductory statistics courses, suggested Shalizi. Thus, while teaching those classes, it would be helpful for educators to clearly articulate the limitations of these assumptions and work to avoid their misapplication in practice. Beyond core statistics-related teaching and curricular improvements, placing greater emphasis on communications training for graduate students could help improve interdisciplinary collaboration between statisticians and domain scientists. Finally, several presenters agreed that the proliferation of complex data and increasing computational demands of statistical inference warrants at least cursory training in efficient computing, coding in languages beyond R,[1] and the basics of database curation.

---

[1] The website for the R project for statistical computing is https://www.r-project.org/, accessed January 4, 2017.

ORGANIZATION OF THIS WORKSHOP PROCEEDINGS

Subsequent chapters of this publication summarize the workshop presentations and discussions largely in chronological order. Chapter 2 provides an overview of the workshop and its underlying goals, Chapter 3 focuses on inference about discoveries based on integration of diverse data sets, Chapter 4 discusses inference about causal discoveries from large observational data, and Chapter 5 describes inference when regularization methods are used to simplify fitting of high-dimensional models. Each chapter corresponds to a key issue identified in the statement of task in Box 1.1, with the second issue of inference about discoveries from data on large networks being interwoven throughout the other chapters.

2

Framing the Workshop

The first session of the workshop provided an overview of its content and structure. Constantine Gatsonis (Brown University and chair of the Committee on Applied and Theoretical Statistics [CATS]) introduced the members of CATS, emphasized the interdisciplinary nature of the committee, and mentioned several recently completed and ongoing CATS activities related to big data, including *Frontiers in Massive Data Analysis* (NRC, 2013) and *Training Students to Extract Value from Big Data: Summary of a Workshop* (NRC, 2014). Alfred Hero (University of Michigan and co-chair of the workshop) said the overarching goals of the workshop were to characterize the barriers that prevent one from drawing reliable inferences from big data and to identify significant research opportunities that could propel multiple fields forward.

PERSPECTIVES FROM STAKEHOLDERS

Michelle Dunn, National Institutes of Health
Nandini Kannan, National Science Foundation
Chaitan Baru, National Science Foundation

Michelle Dunn and Nandini Kannan delivered a joint presentation describing the shared interests and ongoing work between the National Institutes of Health (NIH) and the National Science Foundation (NSF). Dunn said the two agencies share many interests, particularly across the themes of research, training, and collaboration. She described NIH's long history of funding both basic and applied research at the intersection of statistics and biomedical science, beginning with biostatistics and more recently focused on biomedical data science. She introduced the Big Data to Knowledge (BD2K) initiative as a trans-NIH program that aims to address limitations to using biomedical big data. Kannan described NSF's support for foundational research across mathematics, statistics, computer science, and engineering. She noted the broad portfolio of big data research across many scientific fields, including geosciences, social and behavioral sciences, chemistry, biology, and materials science. Since NSF does not typically fund biomedical research, coordination with NIH is important, Kannan said.

Dunn mentioned several NIH programs to improve training and education for all levels, with a focus on graduate and postgraduate researchers—for example, the National Institute of General Medical Sciences Biostatistics Training Grant Program.[1] The BD2K initiative funds biomedical data science training as well as open educational resources and short courses that improve understanding in the broader research community. Kannan described NSF's focus on the training and education of the next generation of science, technology, engineering, and mathematics researchers and educators. She listed examples including postdoctoral and graduate research fellowships, which include mathematics and statistics focus areas, as well as research experiences for undergraduates that can bring new students into the field. Kannan also mentioned the Mathematical Sciences Institutes as an existing opportunity to bring together researchers across many areas of mathematical science, as well as other opportunities for week-long through year-long programs.

Dunn described a third general area of shared interest for NIH and NSF as fostering collaboration between basic scientists typically funded by NSF and the biomedical research community funded by NIH. The NIH-NSF innovation lab provides a 1-week immersive experience each year that brings quantitative scientists and biomedical researchers together to develop outside-the-box solutions to challenging problems such as precision medicine (2015) and mobile health (2016).

Dunn and Kannan said they hoped this workshop would help identify open questions related to inference as well as opportunities to move biomedical and other domain sciences forward. Dunn requested that presenters articulate what biomedical data science research could look like in 10 years and describe why and how it might be an improvement from current practices. Kannan agreed, adding that NSF wants to identify foundational questions and challenges, especially those whose solutions may be applied in other domains as well. She also encouraged speakers to help identify a roadmap forward—not just the state of the art and current challenges, but also what the future holds and what resources are required to get there. Kannan mentioned the National Strategic Computing Initiative (NSCI,

---

[1] The website for the Biostatistics Training Grant Program is https://www.nigms.nih.gov/Training/InstPredoc/Pages/PredocDesc-Biostatistics.aspx, accessed January 4, 2017.

2016) and asked participants to think about what challenges could be addressed with sufficient computational resources.

Chaitan Baru remarked on the rapid growth of data science related conferences, workshops, and events nationally. Similarly, he described the increasing frequency of cross-disciplinary interactions among mathematicians, statisticians, and computer scientists. Both trends were valuable for the emerging discipline of data science, which is bringing together approaches from different disciplines in new and meaningful ways.

Baru described the NSF Big Data Research Initiative that cuts across all directorates. This initiative seeks proposals that break traditional disciplinary boundaries, he said. As NSF spans many scientific domains, a critical objective of the program is to develop generalizable principles or tools that are applicable across disciplines. Across research, education, and infrastructure development, NSF seeks to harness the big data revolution and to make it a top-level priority in the future.

Baru described several high-level challenges that NSF and the emerging discipline of data science are tackling. For example, NSF is seeking to create the infrastructure and institutions that will facilitate hosting and sharing large data sets with the research community, thereby reducing barriers to analysis and allowing easier replication of studies. Regarding education, Baru pointed to the proliferation of master's-level programs but suggested that principles-based undergraduate curricula and doctoral programs are required for data science to become a true discipline. In reference to the White House Computer Science for All program (Smith, 2016), which introduces computing content in high school courses, Baru identified the similar need to introduce data science principles at this level of education.

INTRODUCTION TO THE SCIENTIFIC CONTENT OF THE WORKSHOP

Michael Daniels, University of Texas, Austin

Michael Daniels presented an overview of, and the motivations for, the scientific content of the workshop. He quoted the 2013 National Research Council report *Frontiers in Massive Data Analysis*, which stated, "The challenges for massive data go beyond the storage, indexing, and querying that have been the province of classical database systems . . . and, instead, hinge on the ambitious goal of inference. . . . Statistical rigor is necessary to justify the inferential leap from data to knowledge . . . ." (NRC, 2013). Daniels said it is important to use big data appropriately; given the risk of false discoveries and the concern regarding irreproducible research, it is critical to develop an understanding of the uncertainty associated with any inferences or predictions made from big data.

Daniels introduced three major big data themes that would feature prominently across all workshop presentations: (1) bias remains a major obstacle, (2) quantification of uncertainty is essential, and (3) understanding the strength of evidence in terms of reproducibility is critical. He explained that the workshop was designed to explore scientific inference using big data in four specific contexts:

1. *Causal discoveries from large observational data:* for example, evaluating the causal effect of a specific treatment in a certain population using electronic health records (EHRs) or determining the causal effect of weather on glacial melting using satellite monitoring data;
2. *Discoveries from large networks:* which are increasingly used in biological and social sciences, among other disciplines, to visualize and better understand interactions in complex systems;
3. *Discoveries based on integration of diverse data sets:* for example, combining data from subcellular genomics studies, animal studies, a small clinical trial, and longitudinal studies into one inference question despite each data type having distinct errors, biases, and uncertainties; and
4. *Inference when regularization is used to simplify fitting of high-dimensional models:* specifically how to assess uncertainty and strength of evidence in models with far more parameters (p) than observations (n).

Regarding inference about causal discoveries, Daniels described the tremendous amount of observational data available but noted that this information could be misleading without careful treatment. He emphasized the difference between confirmatory data analysis to answer a targeted question and exploratory analyses to generate hypotheses. He used the example of comparative effectiveness research based on EHRs to call attention to challenges related to missing data and selection bias, confounding bias, choice of covariates to adjust for these biases, and generalizability. Beyond these general challenges, comparative effectiveness research must evaluate the role of effect modifiers and gain an understanding of pathways through which different interventions are acting. Audience member Roderick Little commented that measurement error for big data can be a significant issue that is distinct from bias and warrants attention from statistical analysts.

In conducting inference about discoveries from large networks, the goal is to discover patterns or relationships between interacting components of complex systems, said Daniels. While graph estimation techniques are available, a critical challenge remains in quantifying the uncertainty associated with the estimated graph features and implied interactions, particularly given the high risk of false positives related to big data. Other open questions include development of statistical tests of significance and modification of techniques to analyze dynamic networks that have structural changes over time.

Making inferences based on the integration of diverse data sets poses many of the same challenges—for example, related to missing data and bias in available data—as well as the additional hurdle of integrating data across many different temporal and spatial scales. As an illustrative example, Daniels encouraged participants to think about the challenges and assumptions necessary to estimate the health impacts of air pollution by combining large-scale weather data from satellite images, regional weather stations, localized pollution monitors, and health records.

Analyses of big data often require models with many more parameters (p) than there are observations (n), and a growing number of regularization tools have emerged (e.g., Lockhart et al., 2014; Mukherjee et al., 2015) based on the assumption of sparsity. Daniels explained that the general strategy with these regularization methods is to find the relationships with the greatest magnitude and assume that all others are negligible. While some regularization methods and associated penalties are more helpful than others, there is little formal treatment of uncertainty when these methods are used. This remains an open challenge, according to Daniels. Additionally, many of the current approaches have been developed for relatively simple settings, and it is unclear how these can be modified for more complex systems, particularly when the assumption of sparsity may not be valid. Daniels concluded by stating that because existing statistical tools are in many cases inadequate for supporting inference from big data, this workshop was designed to demonstrate the state of the art today and point to critical research opportunities over the next 10 years.

3

Inference About Discoveries Based on Integration of Diverse Data Sets

The first session of the workshop focused on inference about discoveries from integration of diverse data sets. The session highlighted opportunities and challenges for reliably combining different data types, such as genomic and proteomic data, physiological measurements, behavioral observations, and cognitive assessments, in the context of sound statistical modeling. Alfred Hero (University of Michigan) described the possibilities and risks of big data integration using a case study of genetic biomarker discovery for viral infection and presented a method for uncertainty estimation in graphical network models. Andrew Nobel (University of North Carolina, Chapel Hill) discussed the relationship between statistical and data management challenges when working with large, diverse data sets and presented an iterative testing procedure for community detection and differential correlation mining. Genevera Allen (Rice University and Baylor College of Medicine) discussed two large longitudinal medical cohort studies and described a flexible framework for modeling multivariate distributions with exponential families. Last, Jeffrey Morris (MD Anderson Cancer Center) discussed incorporating biological knowledge into statistical model development to reduce the parameter space and increase the biological coherence of results.

DATA INTEGRATION WITH DIVERSE DATA SETS

Alfred Hero III, University of Michigan

Alfred Hero began by describing the recent Federal Trade Commission report titled *Big Data: A Tool for Inclusion or Exclusion? Understanding the Issues*, which examined some of the potential benefits and pitfalls in drawing inferences from big data (FTC, 2016). While the case studies presented in that report demonstrate a clear commercial value for big data analysis, Hero issued the caveat that "bigger data does not necessarily lead to better inferences" for a number of reasons related to unknown biases, unquantified uncertainty, and uncontrolled variability across studies. In part this is because a lot of big data is collected opportunistically instead of through randomized experiments or probability samples designed specifically for the inference task at hand. Some analyses may have high computational demands and long run times, and Hero pointed to a need for tools that can identify the necessary conditions for replicability—for example, with regard to sample size relative to the number of variables—*before* running the analysis.

Hero described the potential benefits of integrating diverse data sets, including the development of better predictors and better descriptive models. However, realizing these benefits is difficult because assessment of bias and replicability is challenging, especially in high-dimensional cases, and may require more sophisticated methods. Hero described three distinct categories of data integration:

1. *Integration of data within a single study*, in which the analyst has control over the experimental design and all data collection. In principle, bias can be measured, heterogeneity and variability in the data—for example, from batch effects or sample misregistration—can be controlled for, and uncertainty can be quantified. While there will always be some noise in the data due to biological diversity and temporal progression, these studies are the best-case scenario for analysts, albeit expensive and not always feasible. Examples of this type of integration can be found in Wang et al. (2009), Chen et al. (2012), and Hsiao et al. (2014).

2. *Integration of primary data across several studies*, in which the analyst has access to primary data but does not control all elements of experimental design. In this context, the bias and uncertainty in the data are at least partially unknown; for example, there may be one type of data collected by different laboratories with different protocols. This requires a different set of tools for analysts to account for these uncontrolled noise errors—for example, as presented by Deng et al. (2009), Morris et al. (2012), and Sripada et al. (2014).

3. *Integration of metadata across several studies*, in which the analyst does not have access to primary data but instead combines metrics such as mean aggregated effect sizes, computed p-values, or imputed relationships. Examples of this type of post-experimental integration can be found in Singh et al. (2008), Langfelder et al. (2013), and Rau et al. (2014).

Across all of these categories, Hero said there are statistical principles for data integration. Given two data sets X and Y, and assuming a model $f(X, Y | \theta)$ that gives joint probability distribution for X and Y conditioned on the parameter of interest θ, Fisher's sufficiency principle yields that there is a minimal sufficient statistic $T(X,Y)$ satisfying Fisher-Neyman Factorization following equation 1:

$$f(X,Y \mid \theta) = g_\theta(T(X,Y))h(X,Y). \qquad \text{eq. 1}$$

Hero explained that this provides the best possible integration algorithm available and supports any type of inference task, but it requires a model. Furthermore, if the analyst has access to reliable prior information on the variable of interest, the Bayesian posterior distribution induces dimensionality reduction. While these principles provide an optimistic target, Hero reminded the audience that it is challenging to develop such a model and repeated the famous George Box mantra: "all [statistical] models are wrong, but some are useful" (Box, 1979).

There are also practical challenges for using big data, such as the tremendous increase in the amount of information stored on the cloud, said Hero. Looking forward over the next 10 years, Hero described the potential for similar increases in cloud computing and local sharing. As data sets become too large to store and manipulate on a personal computer, questions arise about how to do inference without ever having access to a complete data set (Wainwright and Jordan, 2008; Meng et al., 2013). Similarly, Hero anticipates that privacy concerns—for example, with regard to electronic health records (EHRs)—will result in data sets with more messiness and missing data points as patients opt out of sharing information for research. Approaches for incorporating privacy as a constraint on statistical inference are still in their infancy, said Hero (see e.g., Duchi et al., 2014; Song et al., 2015).

Hero then presented several case studies, the first trying to predict the onset of illness before peak expression of symptoms by looking at various biomarkers over time and identifying the biomarkers that are most useful for predicting the onset of symptoms. Pre- and post-inoculation data describing ribonucleic acid (RNA) expression; protein expression; nasal, breath, and urine cytokines; and self-reported symptoms were collected 3 times daily for 121 subjects over 5 days, resulting in more than 18,000 samples assayed. The collected data have high dimensionality, exhibit large biological and technical variability, and have missing samples. These complexities are typical of big data. Hero presented an overview of select findings from the study, including a novel factor analysis method (Huang et al., 2011; Woods et al., 2013), identification and validation of a biological predictor of viral symptom onset (Zaas et al., 2013), demonstration that use of a personalized baseline reference sample improves predictor performance (Liu et al., 2016), and demonstration that whole blood messenger RNA (mRNA) is the best data type (a "modality") for predicting illness. This study raised additional questions such as whether additional baseline samples could further improve the accuracy of the predictor and how generalizable findings are to the broader population. These questions are currently being explored in a follow-up study that includes more baseline samples and a collection of additional data modalities, said Hero.

These projects are very challenging in terms of both data management and statistical methods, and Hero briefly introduced several inference-driven data integration techniques developed in this project. The novel factor analysis method used to estimate a small number of explanatory variables as biological predictors for onset of symptoms is easily extendable to many data modalities. The novel aspect of this method, Hero explained, is that the positive sum-to-one constraint in the factor model avoids known problems of masking and interference faced by principal component analysis. The novel factor analysis method was more effective for predicting onset of symptoms than other methods in the literature and was validated with additional data sets (Huang et al., 2011; Bazot et al., 2013).

Hero next described the use of gene set enrichment analysis, which integrates variables in the large data sets into known molecular pathways of action (Irizarry et al., 2009) and reduces the dimensionality of the data. Calculating p-values on the differential expression of specific molecular pathways over time allows identification of statistically significant differences. Hero showed a network constructed from correlating these p-value trajectories that groups pathways that have a similar temporal progression of differential expression (Huang et al., 2011), as shown in Figure 3.1.

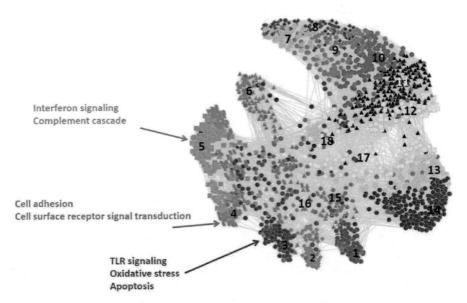

FIGURE 3.1 Spectral clustering of p-value trajectories classifies pathways having similar patterns of differential expression. SOURCE: Reproduced with permission from Huang (2011).

In the final section of his talk, Hero focused on network inference and how to evaluate the significance and replicability of network edges by controlling for false discoveries. This is challenging in the high-dimensional context of big data when fitting models with many more parameters (p) than samples (n), often denoted as $p>>n$, as the classical central limit theorem applies only for fixed p as n goes to infinity. Similarly, Hero described analogous approximations used in the mixed high-dimensional setting that allow both p and n to go to infinity (Bühlmann and van de Geer, 2011), which are not useful for small samples. Hero derived a "purely high-dimensional regime," allowing p to go to infinity with a fixed n (Hero and Rajaratnam, 2011, 2012, 2016), that he used to calculate a critical phase transition threshold (ρ_c) for sample correlation values (equation 2), below which false discoveries dominate:

$$\rho_c = \sqrt{1 - c_n (p-1)^{-2/(n-4)}}. \qquad \text{eq. 2}$$

The phase transition threshold for detection of correlations increases logarithmically with p; thus, for a fixed threshold value one can accommodate exponentially larger p with small increases in sample size (n) in what Hero called "the blessing of high dimensionality." He noted, however, that other inference tasks—for example, full uncertainty quantification—are more demanding in terms of sample size. Hero concluded by emphasizing the importance of rightsizing the inference task to available data by first detecting those network nodes with edges and prioritizing them for further data collection and estimation (Firouzi et al., 2017).

DATA INTEGRATION AND ITERATIVE TESTING

Andrew Nobel, University of North Carolina,Chapel Hill

Andrew Nobel began by remarking that the trade-off between computational error and statistical error is more important than ever because of the growing size of data sets. He cautioned that having a lot of data does not necessarily mean one has the right data to answer the analysis question of interest. Often, the obvious statistical techniques, or those deemed appropriate by a disciplinary scientist, are not the ones actually required. For these reasons, in practice most data analysis problems require sustained interaction between statisticians and disciplinary scientists; the process works best if scientists understand some elements of the statistical analysis, and likewise statisticians understand some of the underlying science.

Nobel then presented a taxonomy of data integration, contrasting classical approaches with the challenges of big data. Classical data integration focuses on integrating data from multiple experiments or multiple groups of samples on a common measurement platform, said Nobel. While this is still relevant today, integration techniques increasingly bring together many data types into a common, or at least overlapping, sample group. Nobel presented an example of a data set from The Cancer Genome Atlas (TCGA) consortium, which contains gene expression data (approximately18,000 genes), micro RNA data (650 miRNAs), copy number data (200,000 probes), methylation data (22,000 regions), and gene mutation data (12,000 genes), as well as patient survival or treatment response descriptors for 350 breast cancer tumors. In such large data sets, frequently data are missing and the analyst does not have coverage across all data modalities for all subjects. For example, Nobel described data from the Genotype-Tissue Expression (GTEx) consortium in which genotype single nucleotide polymorphism (SNP) information is available at 5 million genomic locations for each individual, but gene expression data are available only for a subset of tissues that varies from individual to individual.

The potential benefits of integrating data across many measurement platforms are enhanced statistical power and improved prediction, said Nobel, and these can be used to provide new or better insights into complex biological phenomena. However, this comes at the expense of greater challenges in managing, curating, and analyzing data. Before even getting to formal statistical analysis, Nobel explained, many preprocessing decisions are made—for example, normalization, imputation, and removal of appropriate covariates—that must be scrutinized closely. Even seemingly small decisions may have significant impacts on later analyses, and it can take months to fully understand what processing has been or should be performed on the data. This is another reason for stronger and more frequent collaboration between statisticians and disciplinary scientists, Nobel urged. Moving to model selection, he said that integration of diverse data sets often requires modeling assumptions to reduce the dimensionality of the parameter space—such as sparsity assumptions. While it is critically important to check the validity of these assumptions, this is often more difficult than simply checking for normality. Furthermore, many statistical models have free parameters that must be specified by the analyst, and these decisions also have a significant impact on the final analysis.

Shifting topics, Nobel remarked that networks have become extremely popular in part because they are intuitive visual representations of systems characterized by pairwise relationships and can be amenable to statistical analysis. Unfortunately, networks do not capture higher-order interactions between groups of variables, said Nobel, and summary edge weights (e.g., derived from correlations) may not adequately capture heterogeneity among samples.

He described an iterative testing method for community detection in large networks, in which a community is a group of nodes that are significantly interconnected but have relatively few connections to nodes outside the community (Figure 3.2). Given a group of nodes that represent a candidate community (B_t), the iterative testing algorithm calculates a p-value for each node in the network relative to the configuration null model, which represents how significantly each node is

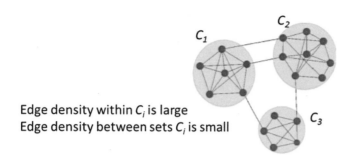

Edge density within C_i is large
Edge density between sets C_i is small

FIGURE 3.2 Community detection in networks identifies sets of nodes that are highly interconnected but have relatively few connections with nodes outside the set. SOURCE: Reproduced with permission from Wilson et al. (2014).

connected to the community B_t. The nodes are ordered from most to least significantly connected to B_t and those nodes with a p-value above a specified threshold are omitted, following the Benjamini-Hochberg step-up procedure (Benjamini and Hochberg, 1995). This process is repeated with this new community (B_{t+1}) and repeats until $B_{t+1} = B_t$. This procedure is competitive with other community detection methods in the literature, is relatively insensitive to the selection of the rejection threshold, and does not require partitioning of the network, said Nobel. Running this process with initial node sets derived from vertex neighborhoods can identify meaningful communities in a network. Importantly, nodes not assigned to any community can be assigned to background.

Nobel described a similar procedure to identify variable sets that are differentially correlated in two sample groups, illustrating the method using cancer subtypes from the TCGA data set. Identifying differentially correlated variables is a useful form of exploratory analysis to generate hypotheses worthy of further study, said Nobel. As a second example, he presented results from differential correlation mining of a functional magnetic resonance imaging (fMRI) data set from the Human Connectome Project. Figures from the analysis show that brain voxels exhibiting differential correlation between language and motor tasks exhibit a distinct spatial pattern that corresponds to known regions of neural activity and that differs from a standard analysis of differential activity. This calls attention to the potential advantages of studying higher-order differences to gain additional biological insights.

PANEL DISCUSSION

A panel discussion followed the presentations from Hero and Nobel. An audience member asked Hero if the gene pathways identified in the viral infection biomarker discovery study comport with known biological mechanisms as well as how or if find-ings were replicated in any of the presented studies. Hero responded that many of the gene signatures identified—toll-like receptor pathways, inflammation pathways, and interferon inducible transcription factors—are well known in the literature. But there are also some factors in this signature that are unknown and seem unrelated to the infection, and the current hypothesis is that these genes and proteins are associated with susceptibility factors that are not currently characterized. Further study may provide some mechanistic understanding of their presence, said Hero. Regarding replication, there have been a number of cross-validation efforts within and across multiple studies, as well as in clinical trials. Interestingly, the biomarker is equally effective for predicting onset of other viral infections, and follow-up research may identify a similar marker for bacterial illness or develop a composite biomarker that can help distinguish the two different causes. Nobel added that interpretation of gene pathways is challenging but often the best available explanation, and Hero agreed, saying that a critical limitation is that gene pathway data present a snapshot rather than account for temporal changes in expression.

Another audience member noted that relying on prediction for model valida-tion provides only an aggregate indicator of model suitability and asked Hero if model residuals were used as an alternative approach to validation. Hero agreed that looking at model fit is feasible, but there is limited value with so few subjects and so many parameters; he would be concerned with overfitting of residuals. He suggested that there is a need to develop methods for residual analysis with few samples and many parameters.

An online participant asked if model development, analysis, and results might be impacted by the fact that much of the data was collected years ago and there are likely differences and improvements in data curation and storage practices over time. Nobel answered first, saying that in his experience there is no such thing as a final data set; rather, the data used in analyses are always evolving. Furthermore, reevaluating any decision about a data preprocessing step can substantially change the downstream analysis. Data preprocessing was discussed extensively in the initial stages of the GTEx project: "As a theoretician . . . I never would have imagined how complex [data curation and sharing] is . . . and I don't know if anyone has the right system," Nobel said. Hero agreed, adding that improvements have been driven to some extent by top journals and funding agencies that require data sharing. None-theless, there is still a long way to go before all data produced from publicly funded research will be available to the public. He noted that large data sets produced by private companies are almost never made public. Hero emphasized that software should be shared more frequently, and Nobel commented that pseudocode is not a substitute.

A member of the audience called attention to the importance of data pre-processing as a crucial part of statistical analysis that should not be viewed as a separate, upstream process. The participant shared some concern regarding the use of p-values in network community detection, because p-values do not measure the size of an effect, particularly with heterogeneous data sets. Nobel agreed, saying that in this specific case the methodological use of p-values was principled, and p-values were not assigned to communities per se (Wilson et al., 2014). The participant said that, in principle, Bayesian approaches can handle many of the challenges raised in this session—for example, missing data or identification problems—but the practice of developing algorithms and computing results is not a trivial matter. The participant described regularization and parameter reduction as a challenge in finding appropriate priors and called for more research into developing Bayesian models for large, complex data structures for which a flat prior is insufficient. Hero agreed, but he remains open-minded and will use whatever method works, whether Bayesian or not.

STATISTICAL DATA INTEGRATION FOR LARGE-SCALE MULTIMODAL MEDICAL STUDIES

Genevera Allen, Rice University and Baylor College of Medicine

Genevera Allen provided an overview of data integration for large-scale multimodal medical studies. Large-scale medical cohort studies typically have many types of data—including clinical evaluations; EHRs; images; gene and protein expression; and social, behavioral, and environmental information—and the objective of data integration in this context is to combine all of these to better understand complex diseases. Allen defined multimodal data as coming from multiple sources or measurement technologies applied to a common set of subjects. If these data take different forms, including discrete, continuous, and binary, they are called mixed multimodal data and require additional processing to integrate. She suggested that multimodal data integration be thought of as the opposite of classical meta-analysis, in which analysts aggregate across many sets of subjects (n) to conduct inference on variables. Conversely, the focus of data integration is to aggregate multiple sets of variables to conduct inference on the subjects.

Allen described some of the data collected through TCGA, which contains information from more than 11,000 patients and totals nearly 2.5 petabytes of data. Genetic data collected include mutation information, copy number variations, gene and miRNA expression, and methylation data. Individually, each modality provides information on only one component of a complex process, and many of the biological malfunctions in cancer development happen between these types of data. Thus, developing a comprehensive understanding of cancer biology requires integration of these diverse data, said Allen. However, integration of these data is particularly challenging both because they are large and because of the mixed data types with different scales and domains. Allen described some of the objectives for integration of TCGA data: (1) to discover the sets of mutations, gene expression changes, and epigenetic changes that are most associated with, and potentially causative of, tumor cell growth; (2) to identify different cancer subtypes and patient groups that have tumors with similar characteristics; and (3) to discover new, potentially personalized, therapies for treating cancer.

In a second case study, Allen introduced the Religious Orders Study (ROS) and the Rush Memory and Aging Project (MAP), which are longitudinal cohort studies tracking dementia and Alzheimer's disease in nearly 3,000 patients. Alzheimer's is a large and growing public health burden, said Allen, and is the only "top-10" cause of death in the United States with no prevention or cure. One key facet of these studies is that patients' brains were donated at death, allowing for full pathological analysis, in addition to containing baseline data describing patients' genetics, lifestyle, environment, and behavior, as well as clinical evaluation, results from 19 distinct cognitive tests, and neuroimaging data collected through the duration of the study. At the time of death, more than 60 percent of patients' brains had symptoms of Alzheimer's disease but less than half of these patients had been diagnosed with the disease, pointing to a large discrepancy between clinical interpretation of cognition and the pathology. Key goals of data integration in these studies include developing a biological understanding of this discrepancy and identifying additional risk factors to guide development of new treatments. Alzheimer's disease is complex, and individual data modalities have been studied extensively, said Allen, making integration an important strategy for advancing understanding of the disease.

After providing background on these two case studies, Allen described some of the practical data-related challenges faced regardless of the statistical methods applied. One critical challenge is identifying and correcting for batch effects, which arise from differences in data collection procedures across different laboratories or technologies and are problematic because they can be confounded across data modalities or across time. Allen showed the results of principal component analysis for methylation data and RNA sequencing data from the ROS and MAP studies, which showed clear grouping and batch effects, in one case due to instrument maintenance. Similarly, structural neuroimaging data collected in the ROS and MAP studies before 2012 relied on a 1.5 tesla magnet that was replaced with a 3 tesla magnet to provide greater resolution; how to reconcile these two batches of imaging data remains an open question, said Allen. Another critical challenge is that not all data modalities are measured for every subject, which creates missing or misaligned data and can result in a very limited sample size if the analysis is restricted only to patients for whom complete data are available (Figure 3.3A). This is problematic because these studies begin with small sample sizes relative to the number of variables. Allen showed several Venn diagrams illustrating the small

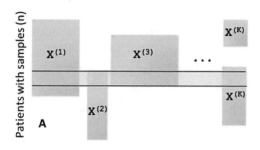
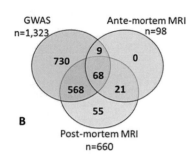

FIGURE 3.3 (A) Generic visualization of how making inferences from the integration of multiple data modalities ($X^{(1)}$ through $X^{(K)}$) can be challenging due to potentially small numbers of patients for whom complete data are available, and (B) illustration with data available from the ROS and MAP studies. SOURCE: Genevera Allen, Rice University and Baylor College of Medicine, "Statistical data integration for large-scale multimodal medical studies," presentation to the workshop, June 8, 2016.

fraction of patients who have multiple data types available: only 68 out of nearly 3,000 patients have genetic information, pre-mortem magnetic resonance imaging (MRI), and postmortem MRI available in the ROS and MAP studies (Figure 3.3B).

Moving to methodological challenges faced in analysis of mixed multimodal data sets, Allen said that prediction is a relatively easy task. Although understudied, there are a number of methods that can be applied: (1) black box methods (e.g., random forests), (2) ensemble learning in which a model is fit to each data modality and then combined, and (3) feature learning approaches (e.g., principal component analysis) on each data modality followed by supervised models using the identified features. More challenging are data-driven discoveries that provide new biological knowledge; for instance, the ROS and MAP studies not only aim to predict if a patient will get Alzheimer's, but also seek to know why, said Allen.

She then discussed a novel method for integrating mixed multimodal data using network models and exponential families to model a joint multivariate distribution. While the broader literature describes network methods for some types of data—for example, the Gaussian network model for continuous valued data and the Ising model for binary valued data—others are less well researched (e.g., count-valued data, bounded random variables), and bringing all of these data types into a common network model is a significant challenge. Allen introduced a framework for graphical models via exponential families, which assumes that all conditional distributions can be described as a distribution from the exponential family (containing many of the most common distributions including the normal, Gaussian, Bernoulli, and Poisson) and takes the general form of the following:

$$P(X) = \exp\left(\theta B(X) + C(X) - D(\theta)\right), \qquad \text{eq. 3}$$

where θ is the canonical parameter, $B(X)$ is the sufficient statistic, $C(X)$ is the base measure, and $D(\theta)$ is the log-partition function. Based on this assumption, it follows that the joint multivariate distribution takes the form of the following:

$$P(X) = \exp\{\Sigma_s\theta_s B(X_s) + \Sigma_s\Sigma_t\theta_{st}B(X_s)B(X_t) + \Sigma_s\Sigma_{t_2\dots t_k}\theta_{s\dots t_k}B(X_s)\Pi^k_{j=2}B(X_{tj}) + \Sigma_s C(X_s) - A(\theta)\},$$ eq. 4

in which the pairwise dependencies between variables s and t are given by the product of their sufficient statistics. This flexible framework accommodates diverse data types, permits a wide range of dependence structures, and allows model fitting with penalized generalized linear models, said Allen. To the best of her knowledge, this is the first multivariate distribution that parameterizes dependencies of mixed data types. According to Allen, one of the main challenges in fitting these models is that each type of data is on a different scale, which requires different levels of regularization, and preliminary work shows that standardization of the different data types is inappropriate and potentially misleading. Furthermore, because of correlation within and between data modalities, there may be confounding interference that obscures weaker signals.

While Allen focused on methods for mixed multimodal data, she called attention to existing work on data integration and dimension reduction for single data types such as joint and individual variation explained (Lock et al., 2013) and integrative clustering (Shen et al., 2009). Developing statistical methods for dimension reduction and clustering within mixed multimodal data sets remains an open problem, said Allen. Referring back to the ROS and MAP studies, she said longitudinal studies with mixed multimodal data also present open statistical problems related to aligning data collected at different times. In the bigger picture, the ROS and MAP studies are two of many ongoing projects looking at aging and cognitive health, which creates the opportunity for meta-analysis across similar integrative studies to increase statistical power, which is an objective of the Accelerating Medicines Partnership–Alzheimer's Disease project.

More fundamental challenges relevant for all medical studies using big data are related to data access, data quality, and patient privacy while sharing data. For example, much of the data collected in the ROS and MAP studies is not yet publicly available, so just getting data to analysts is a major challenge that cannot be overlooked, said Allen. Ensuring reproducibility of research is a critical challenge that is exacerbated in the context of multimodal data because each measurement technique typically has a distinct data preprocessing pipeline, and there may not be one person who understands all of these for large studies such as ROS and MAP. These studies typically are conducted by large teams that use different instruments and different software that introduce error, Allen said, so researchers need to make sure that downstream inferences are reproducible at the end of the analysis.

DISCUSSION OF STATISTICAL INTEGRATION
FOR MEDICAL AND HEALTH STUDIES

Jeffrey S. Morris, MD Anderson Cancer Center

Jeffrey Morris remarked that the last 10 years have produced a large amount of complex, information-rich data that has transformed biomedical research. For example, research in molecular biology produces genome-wide data through a variety of platforms that provides information on deoxyribonucleic acid (DNA), RNA, protein expression, and epigenetics data on mediating processes such as methylation. Similarly, imaging technologies continue to evolve and currently provide both structural and functional information, said Morris. Given the abundance of data, the critical question is how biological knowledge can be extracted from these large, complex, and heterogeneous data sets. Integrative modeling is one of the key scientific challenges and will be critical for translating information to knowledge, said Morris, who showed an example of five different types of neuroimaging that each contains different types of information. This is particularly challenging when data and measurement platforms describe biological phenomena across several orders of magnitude in spatial and temporal scales, ranging from single neurons to whole brain regions.

Integrative modeling faces numerous challenges—for example, the small sample size for complete data sets—said Morris, referencing the small overlapping area in a Venn diagram in Figure 3.3B. This requires implementing creative ways to best use the information that is available or developing strategies for multiple imputation of missing data. Similarly, batch effects are a serious problem if they are confounded with important factors and are even more challenging with complex, high-dimensional data sets. Morris also mentioned the importance of understanding data preprocessing—echoing the call for analysts to consider preprocessing as part of their statistical inference—as well as the practical challenges of storing, indexing, linking, and sharing large data sets. Morris then introduced the issue of defining a common unit of analysis, as different data observe different objects and phenomena. For example, methylation occurs at individual sites on a gene, which can contain more than 20,000 potential binding sites, so it is not trivial to align observations and elements across different platforms.

Morris summarized three main classes of statistical tasks described in the preceding presentation by Genevera Allen:

1. Building of predictive models that integrate diverse data and allow a larger set of possible predictors to search over, which is difficult with mixed multimodal data sets;
2. Structure learning to empirically estimate interrelationships and exploit correlations to reduce the number of parameters, which can be difficult

because although structure can usually be found, not all structures are replicable and meaningful; and

3. Network structure learning and use of graphical models to estimate pairwise associations in the data.

These strategies focus on concatenating across variables to facilitate discovery of underlying biological phenomena, said Morris. However, there are other integrative modeling strategies to narrow the parameter space, specifically incorporating known or theoretical knowledge into model development. For example, genetic information is transcribed to proteins through mRNA, which in turn is modified through miRNA and epigenetics that also have a genetic component. Expression of different proteins results in different phenotypes or molecular subtypes and ultimately causes different clinical outcomes, so it would make sense to build a model that has this type of directed flow between data modalities. Similarly, incorporating biological knowledge from the broader literature, such as molecular pathway information, can inform model development. Additionally, focusing on biologically relevant information—for example, excluding methylation sites that are known not to affect gene expression—can simplify complex data sets.

For the remainder of his presentation, Morris described a case study evaluating subtypes of colorectal cancer that demonstrates incorporation of biological knowledge into integrative modeling. As motivation, he presented a continuum of precision medicine ranging from traditional practices in which all patients are given the same treatment regardless of personal variability to personalized care in which each patient is given a specifically designed treatment. A reasonable middle ground to aim for is identifying and treating cancer subtypes that share many biological characteristics, said Morris. To develop consensus regarding molecular subtypes of colorectal cancer, Morris and colleagues participated in an international consortium led by SAGE Biosystems that combined information from 18 different studies with mRNA data from approximately 4,000 patients. Their analysis yielded four distinct communities, representing different colorectal cancer subtypes, each characterized by different biological characteristics. The subtypes identified are consistently identified and replicable, in part because the data set is large and diverse, said Morris, and he believes it represents true biological differences. Unfortunately, mRNA data are not clinically actionable and require an understanding of the upstream effectors. TCGA data on colorectal cancer and MD Anderson internal data are now being combined with the goal of characterizing the biology of each colorectal cancer subtype to explore questions such as the following: Which subtype has the worst prognosis and is a priority for aggressive treatment? Do different subtypes respond differentially to a specific treatment? Are there subtype-specific targets that can be used to guide development of new treatments? Morris suggested that integrative modeling is critical to answering these questions.

Providing one example, Morris showed boxplots of miRNA expression indicating that the particular miRNA was expressed less in one subtype than in the other subtypes. As miRNA affects expression of numerous other genes, gene set enrichment was used to show that downstream genes that would normally be inhibited by the miRNA were overexpressed in this subtype, and similarly the DNA coding for the miRNA itself is more methylated in this subtype. Putting this information together indicates that methylation inactivates this miRNA, which in turn results in greater downstream expression of genes that are a known hallmark of metastatic cancer (Figure 3.4). Morris described a strategy for relating methylation and mRNA data, which is challenging because methylation is measured at thousands of sites per gene, that involves restricting the analysis to sites for which methylation is correlated with mRNA expression and constructing a gene-level methylation score. In turn, this allowed estimation of the percent of gene expression that is explained by methylation to obtain a list of genes whose expression is strongly modulated by methylation.

In one final example of bringing biological knowledge into statistical model development, Morris presented a Bayesian hierarchical integration framework called iBAG that models biological interrelationships from genetics through clinical outcomes. Beginning with a nonparametric model to regress gene expression based on upstream effectors such as methylation and copy number, these estimates are carried forward as predictors in the clinical regression model. This results in a list of prognostic genes and the upstream effectors that are responsible for its expression, said Morris. The framework has been extended to account for gene pathways as well as to incorporate clinical imaging data, which allows researchers to identify predictive features on an image, understand the gene pathways associated with that feature, and relate these pathways to the upstream genetic and epigenetic processes

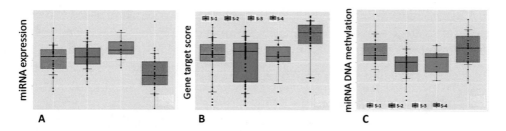

FIGURE 3.4 Combining data from (A) micro RNA expression, (B) gene set enrichment, and (C) gene-level methylation with known biological information allowed researchers to infer that methylation drives differential gene expression in one colorectal cancer subtype. S-1 through S-4 correspond to the different cancer subtypes. SOURCE: Jeffrey Morris, MD Anderson Cancer Center, "Statistical integration for medical/health studies," presentation to the workshop, June 8, 2016.

that dictate gene expression. Integrating multimodal data and known biology reduces the number of potentially relevant parameters, which makes modeling more efficient, and incorporating known biological information yields biologically coherent results that are more likely to be reproducible, said Morris. However, he cautioned that not everything in the literature is true, so incorporating biological knowledge may introduce additional bias. Furthermore, it requires detailed understanding of the underlying biology, and Morris concluded with an appeal for close collaboration between disciplinary scientists and statistical analysts.

PANEL DISCUSSION

Genevera Allen and Jeffrey Morris participated in a panel discussion following their individual presentations. A participant remarked on the growing popularity of graphical models, such as the Ising and Gaussian used by Allen, but asked why Bayesian networks were not mentioned despite their ability to integrate mixed multimodal data and additional advantageous properties. Morris answered first, saying that graphical models can be fit using either a Bayesian or frequentist approach. He agreed that Bayesian networks have many advantages but pointed out that they may be more computationally demanding. Allen responded that a lot of good work has been done using Bayesian networks to integrate mixed multimodal data, and the framework she presented using exponential families to represent a multivariate distribution could be applied with Bayesian networks and priors. Many Bayesian approaches model dependencies between mixed data types in the latent hierarchical structure of the model; this avoids challenges related to scaling of data across modalities but is often more difficult to interpret. There are benefits and drawbacks to both approaches, and data should be analyzed with many different methods, concluded Allen.

Another participant commented that as a biologist he viewed changing technology as a positive trend, making reference to the replacement of 1.5 tesla magnets in MRI machines with a 3 tesla magnet mentioned by Allen, and asked for comments or strategies to avoid older data becoming obsolete while taking advantage of the improved data produced by newer instruments. Allen responded that one strategy is to identify appropriate metrics—for example, gene-level summaries of methylation—that allow researchers to link data sets across technology changes. The strategy depends a lot on the context, Morris explained, but MD Anderson analyzes some common samples with both the newer and older technology and uses these overlapping data sets to create a mapping function to relate the older and newer data sets.

A participant asked how to interpret the dependencies between mixed data specified as multivariate distributions from exponential families and if using likelihood-based methods for inference would lead to model misspecification given the limited sample size. Addressing the first question, Allen responded that dependencies between mixed data are parameterized as the product of sufficient statistics from each underlying distribution and that interpretation of this is an open area for future research. Regarding the second question, Allen agreed that there is an insufficient sample size relative to the number of parameters to rely on likelihood-based inference even in large medical cohort studies with thousands of subjects. In the cases that Allen has applied this approach, she relied on biological knowledge to filter the data, as discussed by Morris, before fitting the network. Furthermore, it is critical to understand the reproducibility of network features, said Allen, who described how she uses bootstrapping to assess the stability of network edges and provides collaborators with a rank ordering of the most important edges. Although this is practicable, Allen acknowledged that there are likely better ways to assess reproducibility that should be the subject of further research.

In the last question, a participant noted that many different models exist and asked how the speakers compared and assessed the suitability of the models they used beyond prediction. Morris answered that after developing a model, he uses it to generate data for comparison to the original data used to create the model. Allen added that model averaging or consensus models are other strategies that can be used to compare and improve inference from complex models.

4

Inference About Causal Discoveries Driven by Large Observational Data

The second session focused on inference about causal discoveries from large observational data such as electronic health records (EHRs). A primary goal of biomedical research with big data is to infer causative factors such as specific exposures or treatments; however, theories of causal inference in observational data (e.g., Pearl, 2000) remain relatively open in the context of large, complex data sets containing many treatment variables with possible interactions. Comparative effectiveness research using EHRs faces challenges related to potentially large amounts of missing data ("missingness") and associated bias, confounding bias, and covariate selection. Joseph Hogan (Brown University) discussed mathematical and statistical modeling of human immunodeficiency virus (HIV) care using EHRs. Elizabeth Stuart (Johns Hopkins University) further elaborated on synergies and trade-offs between mathematical and statistical modeling in the context of causal inference and public health decision making. Sebastien Haneuse (Harvard University) described a comparative effectiveness study on the effects of different antidepressants on weight gain using EHRs. In the last presentation, Dylan Small (University of Pennsylvania) demonstrated isolation of natural experiments within EHRs through a case study evaluating the effect of childbearing on workforce participation.

USING ELECTRONIC HEALTH RECORDS DATA FOR CAUSAL INFERENCES ABOUT THE HUMAN IMMUNODEFICIENCY VIRUS CARE CASCADE

Joseph Hogan, Brown University

Joseph Hogan discussed the use of EHRs to compare recommendations for when to initiate HIV care for patients in low- and middle-income countries. Although the work is still ongoing, this type of analysis can be used to infer how different treatment strategies impact patient progression through the HIV "care cascade." He explained that the HIV care cascade is a conceptual model for understanding progression through stages of HIV care and is composed of (1) diagnosis, (2) linkage to care, (3) engagement or retention in care, (4) prescription of antiretroviral therapy (ART), and (5) viral suppression. Recommendations for when patients who test positive for HIV should begin treatment have evolved over time, said Hogan, in part due to the lack of available ARTs. In 2003, the World Health Organization recommendation was to begin ART if a patient's cluster of differentiation 4 (CD4)—an aggregate measure of immune system health—fell below 200 cells/microliter, and this threshold has continually been increased due to accumulating evidence. The most recent guidelines call for initiating treatment for all HIV-infected individuals, regardless of CD4 count. The HIV care cascade is used to help formulate health care goals and institutional benchmarks—for example, the 90-90-90 goal put forth by the United Nations. This benchmark aspires to have, by 2020, 90 percent of HIV-infected patients aware of their status, 90 percent of those diagnosed with HIV receiving ART, and 90 percent of those receiving therapy having viral suppression, explained Hogan. These benchmarks present concrete objectives and create an impetus for empirical evaluation that must take into account statistical uncertainty, said Hogan. Evaluating progress in reaching these benchmarks is challenging because the necessary data are complex and come from multiple sources, and the care cascade is a dynamic and complex process. Furthermore, there is a trade-off between rigor and clarity, said Hogan, as program managers need interpretable information to evaluate progress and make decisions about new policies or interventions.

One of the simplest approaches is to aggregate data into a histogram showing the number of patients in each stage, which presents a static snapshot and provides limited insight into the cascade as a process. More recently there is growing interest in modeling the entire HIV care cascade, typically using microsimulation techniques based on complex, nonlinear state-space mathematical models. Typically these approaches assume an underlying parametric model, aggregate data from numerous different sources to quantify relevant parameters, calibrate the model against known target outcomes, and then explore the effects of alternative interventions through iterative simulation. Mathematical models are modes of synthesis that are advantageous because they can represent highly complex systems, and calibration ties the model to some observed data, said Hogan. However, there are several limitations to keep in mind when interpreting results; for instance, it is unclear if the models represent one population of interest and reflect causal effects when parameter values are adjusted. Hogan provided an example from the literature of one such model used to compare the effects of different treatment strategies on patients' progression through the care cascade (Ying et al., 2016; Genberg et al., 2016). The model specifies 25 states as a function of a viral load and CD4 count and defines a system of equations to calculate state transition probabilities based on primary data and the literature. With values for all parameters, the model is run and calibrated using 30 years of historical data, and then the model is used to project differences in HIV prevalence and incidence associated with alternative scenarios of treatment and home counseling. Nonetheless, Hogan said, without formal measures of uncertainty it is unclear whether the modeled differences in HIV incidence are significant.

With growing availability of EHRs containing longitudinal data for thousands of patients, Hogan explained that it is possible to develop statistical models of the HIV care cascade that are representative of a well-defined population in actual care settings. However, he cautioned that using observational EHR data can be challenging compared to using data from a cohort study due to irregular observation times and abundance of confounding factors. More fundamentally, it is difficult to operationalize concepts such as "being retained in care" from observational EHR data, said Hogan, and requires up-front effort somewhat analogous to preprocessing of genomics data.

Using data extracted from EHRs and maintained by the Academic Model Providing Access to Healthcare (AMPATH) consortium, Hogan described a statistical model for comparing the effects of two HIV treatment strategies—treat upon enrollment regardless of CD4 count or treat when CD4 ≤ 350cells/microliter—on patient progression through the care cascade. The database included information on over 57,000 individuals in care. A patient can be in one of five well-defined states: engaged, disengaged, lost to follow-up, transferred, or dead (Figure 4.1A). The probability of transitioning from one state to another is calculated from the aggregated AMPATH EHR data set. Based on these probabilities, the one-step transition model is used to predict the probability of state membership from enrollment through 1,800 days (Figure 4.1B), which stops short of causal inference but presents data in an easily interpretable format for health care decision makers.

Hogan extended this model to compare the two treatment strategies mentioned above by assuming that treatment is randomly allocated for individuals sharing the same observed-data history, that the length of follow-up depends only on the observed-data history, and that the model follows first-order Markov dependence. Thus, implementing this method requires fitting a sequence of observed-data

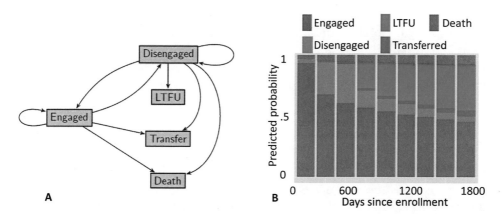

FIGURE 4.1 (A) One-step transition model used to represent patient progression through the HIV care cascade and (B) an illustration of how predicted state probabilities can be represented as a function of time. NOTE: LTFU, lost to follow-up. SOURCE: Joseph Hogan, Brown University, "Using EHR data for causal inferences about the HIV care cascade," presentation to the workshop, June 8, 2016.

models for state membership conditional on previous CD4 count, treatment history, and other covariates, as opposed to the complex model assumed in mathematical models. The method also requires an observed-data model for time-varying covariates such as CD4 count, which can be checked for adequacy and elaborated if the fit is poor. Hogan illustrated how to implement the method for the case where time-varying CD4 count is treated as the main confounder. The method is implemented with a G computation algorithm using Monte Carlo simulation to average over the distribution of time-varying CD4 count (i.e., to calculate the high-dimensional integral over multiple time points). Hogan showed that the resulting predicted-state probabilities over time for the two treatment strategies that indicate starting treatment upon diagnosis and regardless of CD4 count leads to more patients remaining engaged and fewer patients lost to follow-up. Building a statistical model allows for formal quantification of uncertainty and application of standard methods, such as confidence intervals and hypothesis tests, to support inferences about effects of interest.

In comparing statistical and mathematical approaches to modeling the HIV care cascade, Hogan described the former as beginning with as much data as are available and building a simple mathematical model to describe the data, whereas mathematical models typically focus on building a more complex process model and then using select historical data for calibration. In short, mathematical models focus more on the model and less on the data, and conversely statistical models focus more on the data and less on the model. Given the abundance and complexity of information in EHRs, there is an opportunity to combine elements of both approaches to make models that align closely with data but support a higher level of complexity, said Hogan. Mathematical modeling techniques can enrich statistical models in several ways—for example, in representing missing information in data sets or integrating outside data—with the goal of increasing model complexity while maintaining close alignment with observed data from a specific cohort. He suggested that statistical principles be integrated into informatics systems to ensure that decisions based on EHR data are well-grounded by rigorous inferences. Similarly, Hogan described the need for a framework for grading evidence that is used and produced by models—for example, by considering if the data describe a well-defined population of interest, if different sources of uncertainty are identified, if the model fit is evaluated, among other criteria. Hogan concluded that as complex, messy, information-rich EHR data become increasingly available and are potentially used to inform treatment decisions, practice patterns, and health care policy, "Statistical principles could hardly be more important."

DISCUSSION OF CAUSAL INFERENCES ON THE HUMAN IMMUNODEFICIENCY VIRUS CARE CASCADE FROM ELECTRONIC HEALTH RECORDS DATA

Elizabeth Stuart, Johns Hopkins University

In mental health research there is increasing interest in comprehensive systems modeling, said Elizabeth Stuart, which is an area ripe for combining mathematical and statistical modeling. However, large-scale mathematical models typically require more assumptions than statistical models and may contain unnecessary complexity that is irrelevant to the specific decision context. There is a tension between creating large, detailed models that are descriptive of real-world complexities and small, simpler models that are often required to answer a given question. Thus it is critically important to target analyses toward particular parameters or questions of interest, said Stuart, as a way to simplify models and limit assumptions. For example, structural equation models are powerful for exploratory analyses because they include more effects and correlations than other inference techniques such as linear regression; however, these models require additional assumptions that may not be valid or necessary to answer questions related to a single causal effect (VanderWeele, 2012). When using methods that require strong assumptions and data without good empirical estimates, Stuart encouraged researchers not only to acknowledge and communicate the limitations, but also to assess the sensitivity of results to those limitations. It is critical to keep the ultimate goals and quantities of interest in mind when developing methods to evaluate evidence, she said.

Another challenge when estimating and drawing more complex models for causal inference is formalizing the potential outcomes. For example, in comparing the effectiveness of two treatment strategies there are two potential outcomes: a patient either receives treatment or stays in the control group. These are different random variables that are almost never both depicted on graphical models. Stuart said that it can be problematic to bundle the two outcomes together because it is possible that there is a different relationship between the covariates and outcome under one treatment than the other treatment. To demonstrate this point, Stuart described her experience analyzing the effectiveness of a marital intervention using the self-reported variable "relationship quality," which was interpreted differently by subjects in the intervention (treatment) and control groups. For example, some subjects who did not receive the intervention and divorced still reported high relationship quality because they rarely had to interact with their partner, whereas other subjects conceived of "relationship quality" differently.

Regarding standards of evidence and the pros and cons of different study designs, Stuart described the ongoing debate in mental health research regarding experimental versus nonexperimental studies. Conventional wisdom is that experimental studies are always preferable because they allow greater control and lower bias, but this may be misleading when the objective of the study is to estimate a population treatment effect, said Stuart. It is critical to consider both internal and external validity as well as sources of bias (Imai et al., 2008), she said, and when estimating population effects it is possible that a small nonrepresentative randomized trial actually has more bias than a large nonexperimental study in a representative sample. Stuart was optimistic that with increasing access to big data such as population-wide EHRs, there would be the opportunity to conduct well-designed nonexperimental studies that provide better evidence about treatment effectiveness in the real world.

In conclusion, Stuart reiterated that the community needs tools to assess which parameters and assumptions matter most in large, complex models and to prioritize these for further investigation and refinement. One strategy is to formalize sensitivity analyses, particularly in large, complex mathematical models with many underlying assumptions and parameter values that are characterized by high or unknown uncertainty. In addition to uncertainty, there is a need for methods to account for the fact that data drawn from multiple studies represent different populations, said Stuart. Finally, it is relatively easy to perform validation exercises with predictions, but validation of causal inferences is more nuanced and requires further work.

PANEL DISCUSSION

During a panel discussion with Joseph Hogan and Elizabeth Stuart following their presentations, a participant mentioned a technique called Bayesian melding[1] and inquired as to how much it informed Hogan's work. Hogan responded that he is aware of this technique and similar work on Bayesian calibration that specifies full likelihood distributions for variables and then collects them into a common model. Moderator Michael Daniels (University of Texas, Austin) commented that although the methods may be similar, the application contexts are different in that EHRs provide much richer data. Hogan explained that one approach for the EHR context would be to build out from a statistical model—for example, the AMPATH data have limitations in that many patients are lost to follow-up and, despite some evidence from other sources that many of the patients have died, the EHR data have incomplete information on mortality. Researchers associated with AMPATH have done tracing studies to track down a subset of patients who are lost to follow-up and then incorporate that information back into the statistical model to generate new estimates of mortality across the population of interest. Going a step further, said Hogan, would be identifying data collected from other studies to integrate into the AMPATH study, but this would require a strategy to account for the different populations being represented.

Another participant inquired how confounding variables should be identified, mentioning ongoing work in data-driven identification of relevant confounders using large, messy medical claims data sets. Stuart answered that confounding variable selection should ideally be driven by scientific understanding and be independent of the data, and she emphasized the importance of a strong separation between research design and analysis. This is often difficult in an EHR scenario, particularly if the analysis is looking across large numbers of parameters and outcomes. Additionally, much of the data-driven confounder selection work has been done with data from single time points, and it is important to consider how this changes with time-varying confounders. Hogan commented that CD4 was chosen as a confounder because it is arguably the strongest measured predictor of treatment selection and outcome, as evidenced by both empirical and scientific understandings of ART's role in HIV suppression. Regarding the broader question, empirical selection of confounders is another example where there is tension between statistical theory and practice, said Hogan, because in theory it is impossible to know whether a sufficient set of confounders has been selected. He said he is hesitant to apply data-driven approaches to confounder selection using data such as those from an EHR, in part because the data structure is so irregular, but perhaps more importantly because it might not be a representative sample from

---

[1] Bayesian melding combines mathematical modeling with data (e.g., Poole and Raftery, 2000).

a population of interest. Hogan described a hypothetical example: suppose HIV treatment recommendations are based on the CD4 count being above a specified threshold, yet in practice treatment is not allocated strictly according to the policy. That is, many people with CD4 just over the threshold receive treatment while many just under the threshold do not. If patients in a sample are selected from among those whose CD4 is near the threshold, it is possible that CD4 will not be chosen as a potential confounder based on purely empirical grounds because, in this subset, it will not be strongly correlated with receiving treatment. In this case, a person with contextual knowledge would recognize that this sample may not be well suited to addressing the causal question because it does not allow leveraging of important information from a well-known confounder. This limitation cannot be discovered by a confounder selection method that is purely empirical. The fundamental inferential issue in this hypothetical example is not the method used to select confounders, but rather whether analysts have a properly selected sample.

In the final question, a participant commented that mathematical models and microsimulation typically have some calibration against observed data taken to be ground truth and asked what was analogous when using statistical models for causal inference. Stuart responded that for any given individual, the causal effect cannot be known because one potential outcome might be observed while there is no information on the other. In practice, researchers can calibrate the control group and calibrate the treatment group, but there is a missing piece in bringing these together. Hogan agreed, saying that researchers should test sensitivity of findings to departures from key assumptions, particularly when drawing causal inferences from large observational data. Perhaps the most important of these is the assumption of "treatment ignorability" or "no unmeasured confounding." Moreover, correctly predicting the outcome of a treatment from observed data does not validate that the treatment caused that outcome. He encouraged participants to think of causal inference in terms of factoring a joint distribution of observed and unobserved potential outcomes, and he noted that clearly separating these two components in a model makes untestable assumptions clear and leads to more coherent and transparent inferences.

A GENERAL FRAMEWORK FOR SELECTION BIAS DUE TO MISSING DATA IN ELECTRONIC HEALTH RECORDS-BASED RESEARCH

Sebastien Haneuse, Harvard University

Sebastien Haneuse began by reiterating a fundamental difference in the scientific goals of comparative effectiveness research—for example, Hogan's presentation comparing two HIV treatment strategies—and exploratory analyses, as discussed in the workshop's first session. With this context, Haneuse described a case study using EHRs to evaluate whether different antidepressants or classes of anti-depressants have differential impacts on patients' long-term (i.e., after 24 months) weight change. Some drugs were hypothesized to lead to weight gain while others were not, and these side effects were assumed to be independent of effectiveness of the antidepressant. The analysis was a retrospective longitudinal study conducted with electronic databases maintained by the Group Health Cooperative, which contained full EHRs based on EpicCare as of 2005; pharmacy data dating back to 1977; and additional databases that track demographic data, enrollment information, claims data, and primary care visits. Inclusion-exclusion criteria were as follows: (1) the patient must be between 18 and 65 years of age, (2) the patient must have undergone a new treatment between January 2006 and December 2007, and (3) the patient must have been continuously enrolled for at least 9 months. Application of these criteria to the available EHR data resulted in a sample of roughly 9,700 patients from which data on weight, potential confounders, and auxiliary variables were extracted for the 2-year interval prior to initiation of the new treatment through 2009. Although weight is a continuous variable, EHR data contain only snapshots of a patient's weight trajectory, said Haneuse, and there is wide variability among the frequency of visits across patients. Thus, some patients have rich information on weight change, while the majority of patients have limited information.

EHR data can provide a large sample size over a long time period at a low cost compared to dedicated clinical studies; however, it is critical to remember that EHR data were not collected to support specific research tasks. Haneuse encouraged researchers to compare available EHR data to the data that would result from a dedicated study, saying that observational data probably do not have comparable quality and scope. There are additional practical challenges to using EHR data, such as extracting text-based information contained in clinician notes, inaccurately recording information, linking patient records across databases, and confounding bias. Although some of these challenges are not new and are encountered in traditional observational studies, existing methods for addressing them are ill suited for the scale, complexity, and heterogeneity of EHR data, said Haneuse. There is an emerging literature focused on statistical methods for comparative effectiveness research with EHRs that has focused largely on resolving confounding bias, whereas problems of selection bias and missing data are underappreciated. In the context of the antidepressant study, EHR data would ideally contain information on weight at baseline and 24 months for each patient, yet in reality less than 25 percent of patients had both measurements, leaving 75 percent of patients with insufficient information to compute the primary outcome. One approach is to simply restrict the analysis to the subsample of patients for which complete data are available. However, this raises questions about the extent to which conclusions drawn from the subsample are generalizable to the population of interest. Haneuse defined selection bias as the difference between conclusions drawn from the subsample and conclusions that would result from complete information for the entire study population. Selection bias can thus be framed as a missing data challenge, for which there is a large amount of literature and methods that may be useful.

The validity of all methods for missing data relies on the assumption that data are missing at random, which Haneuse explained means that missingness depends solely on known variables, and thus the analyst can control for missingness with known information. Evaluation of this assumption typically proceeds by conceiving of a potential mechanism that determines whether or not data are missing, identifying factors that influence this mechanism, and evaluating whether these factors are described in available covariates. Haneuse argued that focusing on a single mechanism is overly simplistic for EHRs, which are high dimensional and heterogeneous, and in the context of clinical health care provision. For example, in the case of antidepressant-related weight change, there is no single mechanism that is responsible for determining whether a patient's weight is recorded, but rather a complex combination of patient and clinician decisions. This simplistic approach may not fully account for missingness and may result in residual selection bias, which can compromise the generalizability and utility of results, said Haneuse.

Haneuse described a general framework for addressing selection bias in analysis of EHR data. Given the high complexity and heterogeneity of EHR systems, it is unlikely that any single method will be universally applicable, so Haneuse suggested that researchers consider two guiding principles: (1) identify the data that would result from the ideal study designed to answer the primary scientific question and (2) establish the provenance of these data by considering what data are observed and why. This process will generally involve identifying all variables that would have been collected and indicating the timing of all measurements, with additional details depending on the goal of the study, before even looking at available EHR data. Conceiving of the ideal study and resulting data allows for a concrete definition of complete and missing data, which analysts can use to characterize why any given patient has complete or missing data. Specifically, Haneuse presented the general strategy of decomposing the single-mechanism model of why a patient has missing data into a series of more manageable submechanisms, with each submechanism representing a single decision. In the antidepressant-related weight gain case study, Hanuese proposed three sequential submechanisms that each partially determines if a patient's EHR data contain a weight measurement at 24 months: the patient must be enrolled at 24 months, the patient must have been treated at 24 months (±1 month), and the patient's weight must have been recorded at this treatment (Figure 4.2).

Breaking down missingness into more granular submechanisms allows each to be explored in greater detail, whereas the single-mechanism approach represents an average of these mechanisms. Haneuse evaluated the number of patients meeting the criteria of each submechanism, calling attention to the heterogeneity across

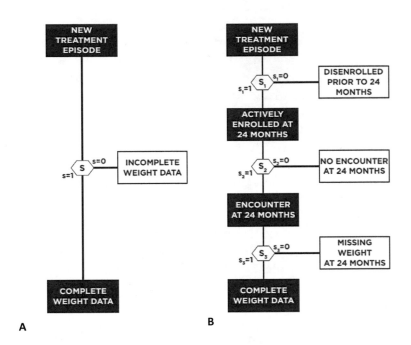

FIGURE 4.2 Comparison of (A) a simple single-mechanism approach to modeling missing data and (B) a more detailed specification consisting of three submechanisms responsible for missingness. NOTE: S_x, random variable; S=0, missing information; S=1, complete information. SOURCE: Reproduced from Haneuse, S., and M. Daniels. 2016. A general framework for considering selection bias in EHR-based studies: What data are observed and why? *eGEMS (Generating Evidence & Methods to Improve Patient Outcomes)* 4(1): article 16. doi:http://dx.doi.org/10.13063/2327-9214.1203, http://repository.edm-forum.org/egems/vol4/iss1/16, licensed under a Creative Commons Attribution-Noncommercial-No Derivative Works 3.0 License (https://creativecommons.org/licenses/by-nc-nd/3.0/).

patient EHRs. Showing odds ratios from logistic regression models for the single-mechanism and three-mechanism models, Haneuse emphasized that different covariates can have different effects on each submechanism, which makes it difficult to interpret the significance of coefficients from the single-mechanism model. While the preceding example focused on three specific mechanisms, Haneuse said there are many alternative submechanisms that cause missing data that could be considered. For example, patients could receive treatment from an outside medical system or could receive clinical advice via the phone or secure messaging, both of which would result in missing weight measurements at 24 months. However, not all mechanisms will be relevant for any given EHR context or analysis question.

The proposed framework enhances transparency in assumptions regarding missing data, facilitates elicitation of factors relevant to each decision, and promotes closer alignment between statistical methods and the complexity of the data, said Haneuse. Existing methods such as inverse probability weighting or multiple imputations can be combined with the multi-mechanism framework proposed, and multiple methods can be mixed in one analysis—for example, use of inverse probability weighting for some submechanisms and imputation for others—in what Haneuse termed "blended analyses." Similarly, data-driven methods for variable selection developed to address bias from confounding may be adapted to the context of selection bias to understand which submechanisms and variables are most relevant. One of the most interesting areas, said Haneuse, is using this approach to prioritize future data collection efforts to supplement the information available in EHRs. He emphasized that these methods will become increasingly important as EHRs become the norm in clinical practice.

Haneuse concluded that EHR systems may be designed for secondary research purposes in the future but, until then, the majority of researchers using EHRs will try to model the best they can with the most information possible. Haneuse suggested that grounding the modeling within the context of the ideal designed study is appealing because it requires explicit definition of both the target population of interest and what it means to have complete data. This approach can help focus the scientific inquiry at the expense of unutilized information (for example, all of the patient weight measurements between 2 and 22 months), but this may be a relatively small price to pay, said Haneuse.

DISCUSSION OF COMPARATIVE EFFECTIVENESS
RESEARCH USING ELECTRONIC HEALTH RECORDS

Dylan Small, University of Pennsylvania

Dylan Small reminded participants that using EHRs for comparative effectiveness research can be cheaper, faster, more representative of real-world effectiveness, and more statistically powerful (due to large sample sizes) than randomized trials. However, comparative effectiveness studies using EHRs face serious challenges regarding confounding and selection bias, said Small. He added that Haneuse presented an elegant framework for addressing selection bias that can better engage clinicians in thinking through why data are missing. Small remarked that another potential application of Haneuse's multi-mechanism framework could be identification of subsets of the data for which missingness could be regarded as missing at random, analogous to selective ignorability for confounding bias (Joffe et al., 2010).

Shifting to confounding bias in EHR data, Small recounted calls to largely replace randomized trials with EHR-based studies (Begley, 2011), because EHRs contained detailed information on sufficient confounding variables to safeguard against errors that beguiled other observational studies in the past. However, Small cautioned that confounding by indication—for example, that clinicians likely prescribe more aggressive treatments to patients with worse-perceived prognoses even if that treatment has a higher likelihood for side effects—will persist in comparative effectiveness studies using EHR data. He described Olli S. Miettinen's warning that when a rational reason for an intervention exists, it tends to constitute a confounder (Miettinen, 1983); thus, the need for randomized trials is accentuated in studying a desired effect (e.g., comparative effectiveness research) more than when studying an unintended effect (e.g., comparative toxicity studies). Such confounding can be quite subtle and complex in the clinical health care context, and comparative effectiveness studies using EHR data need to critically consider why two patients with similar observed covariates received different treatments. Unless the treatments are assigned entirely at random, any causal inference made is subject to bias and is potentially misleading, said Small.

One strategy to develop reliable causal inferences from EHR data is to look for natural experiments contained in subsets of the data and apply quasi-experimental tools, such as multiple control groups or secondary outcomes known to be unaffected by the treatment, to test for hidden bias. Small described one example comparing the effect of childbearing on workforce participation (Angrist and Evans, 1998) to evaluate whether having more children reduces a mother's working hours. He explained that answering this question presents challenges related to confounding by indication because many births are planned and nonrandom (Zubizarreta et al., 2014). One tool to isolate natural experiments from larger data sets is to focus on differential effects by comparing workforce participation between women who had twins versus single births, for example, as opposed to comparing women who have children to those with no children. "Risk set matching," a second tool for isolating natural experiments, compares women who had similar covariates up to the time of birth—for example, comparing women who have had the same number of births (among other variables) as opposed to comparing mothers with three children to first-time mothers. Small showed results from the isolated natural experiment indicating that the median fraction of a 40-hour work week performed was slightly (approximately 8 percent) lower for mothers of twins compared to the control groups, and this difference became more pronounced with increasing numbers of births.

In a second example of comparing differential effects, Small described evaluation of the effects of pain relievers on Alzheimer's risk, testing the theory that non-steroidal anti-inflammatory drugs (NSAIDs) such as Advil™ may reduce patient risk. However, there is potential confounding from the possibility that early-stage Alzheimer's patients are less aware of physical pain and thereby less likely to take NSAIDs. To control for this confounding, Small suggested comparing Alzheimer's risk between patients taking NSAIDs and those taking non-NSAID pain relievers like Tylenol™ as one such natural experiment available in EHR data. Small concluded that EHRs and other sources of big data do hold a lot of promise, but long-known problems of selection and confounding bias must be addressed. Isolating natural experiments is one strategy to reduce confounding bias.

PANEL DISCUSSION

In a follow-on panel discussion with Sebastien Haneuse and Dylan Small, a participant described Haneuse's strategy of comparing EHR data to data that would result from the ideal randomized trial as the appropriate way for statisticians and other researchers to reason through causal discoveries made from big data, as opposed to simply extracting as much as possible using advanced methods such as machine learning. The participant asked if Haneuse had considered a way to quantify the extent of missingness, similar to a missing information ratio, relative to the ideal study design. Haneuse responded that the approach does allow for concrete definition of complete and missing data, though he had not thought about this particular issue of quantification.

Another participant asked generally what researchers in the field could do to raise awareness of these issues and make the appropriate statistical methods available, particularly given the anticipated growth in EHR data collection and availability. Haneuse said that it takes years before statistical methods from literature are adopted by practitioners, and he suggested that researchers engage in similar workshops and discussions more frequently. Another participant said that many issues and sensible approaches to using EHR data are known and documented in the literature, and Small commented that some journals are more statistically rigorous than others.

One participant said that conceiving of the ideal study design as a basis for defining data needs was potentially risky because it could lead one to disregard available data that do not correspond to this design—for example, not using patient weight measurements taken at 18 months. Haneuse clarified that those measurements *do* have useful information that should be used in solving the missing data challenge and reiterated that researchers should use the ideal study design to help define complete and missing data, using any information that is available.

5

Inference When Regularization Is Used to Simplify Fitting of High-Dimensional Models

The third session of the workshop explored inference after a regularization technique is applied to reduce the number of parameters being fit in a model. In many analyses of big data, the number of variables (p) described in available data greatly exceeds the number of observations (n), which presents challenges for classical inference methods based on the assumption that n is much larger than p. Approaches for such high-dimensional inference that rely on sparsity assumptions—that the number of nonzero effects is limited—have emerged over the past two decades. However, important questions remain regarding methods for uncertainty quantification and the validity of underlying assumptions for increasingly complex questions and data sets. Daniela Witten (University of Washington) introduced novel methods for learning the structure of a graphical model from gene expression and neural spike train data, and she discussed the gap between statistical theory and practice in the context of theoretical results associated with high-dimensional model fitting. Michael Kosorok (University of North Carolina, Chapel Hill) elaborated on model selection consistency[1]—the consistency of the support of the selected model—and suggested alternative approaches to inference after regularization, concluding with an appeal to consider the validity of model assumptions and to develop new methods with less stringent assumptions. Jonathan Taylor (Stanford University) introduced methods for selective inference, where data splitting and data carving are used for both model selection and the inference task.

---

[1] For additional information on model selection consistency, see Zhao and Yu (2006) and Lee et al. (2013).

Emery N. Brown (Massachusetts Institute of Technology, Massachusetts General Hospital, and Harvard Medical School) reviewed the Box-Tukey paradigm in the context of neuroscience and discussed the importance of statistics education broadly. Xihong Lin (Harvard University) described the growing size of genetic and genomic data sets and associated statistical challenges and emphasized the importance of statisticians engaging early in experimental design and data collection.

LEARNING FROM TIME

Daniela Witten, University of Washington

Daniela Witten began by describing methods for learning the structure of graphical models, which represent interrelationships between multiple random variables (Figure 5.1A), from large biomedical data sets that contain measurements taken over time. Time is incredibly important in biological processes, said Witten, providing the illustrative example of progenitor cells developing through several stages to become mature muscle fibers. If available data only represent observations from one point in time, it is challenging to develop a complete understanding of this dynamic biological process. Alternatively, averaging data representative of these different stages over time may also be misleading and may potentially undermine the goals of the inference task.

Witten presented a simplified example using gene expression data collected at discrete time points for nine genes—in real usage scenarios, there are typically thousands—with the goal of creating graphical models to represent regulatory

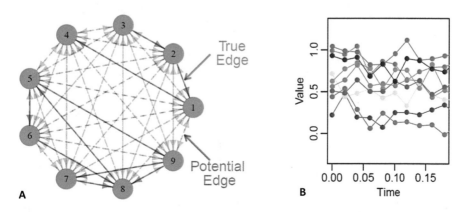

FIGURE 5.1 Example of a (A) directed graphical model with nine nodes, constructed from (B) gene expression time course data. SOURCE: Produced by Shizhe Chen and presented by Daniela Witten, University of Washington, "Learning from time," presentation to the workshop, June 9, 2016.

relationships between individual or groups of genes (Figure 5.1B). She proposed a model in which the observed time-course data (Y_j) for expression of each of p genes can be modeled as a smooth noiseless trajectory (X_j) that cannot be directly observed, plus an error term (ε_j):

$$Y_j(t_i) = X_j(t_i) + \varepsilon_j(t_i). \qquad \text{eq. 5}$$

Witten described an additive differential equation model for the underlying noiseless trajectories as follows:

$$\frac{d}{dt}X_j(t) = C_j + \sum_{k=1}^{p} f_{jk}\left(X_k(t)\right), \qquad \text{eq. 6}$$

where the function f_{jk} is unknown. Witten explained that it is not necessary to know the functional form of f_{jk} for structure learning beyond knowing whether it is exactly zero, which indicates that there is no regulatory relationship or corresponding graph edge. Fitting this nonparametric model is challenging, explained Witten. As f_{jk} is unknown, she described approximating it with a set of fixed basis functions $\psi = (\psi_1 \ldots \psi_M)^T$ (Ravikumar et al., 2009) such that

$$\frac{d}{dt}X_j(t) \approx C_j + \sum_{k=1}^{p} \psi\left(X_k(t)\right)^T \theta_{jk}. \qquad \text{eq. 7}$$

Equation 7 is linear in θ_{jk}, a vector of length M, which simplifies model fitting significantly. However, the high dimensionality of the model is still challenging, with order Mp^2 unknown parameters, and the number of time points at which there are measurements (N) is typically much smaller than Mp^2. To reduce the dimensionality of the problem, Witten applied a group lasso penalty to induce sparsity in the data by encouraging the vector θ_{jk} to be either exactly zero or completely nonzero (Yuan and Lin, 2006; Simon and Tibshirani, 2012).

Another critical challenge faced in fitting the model shown in equation 7 is that the underlying noiseless trajectory $X_j(t)$ for the expression of gene k is unobserved, said Witten. In practice, analysts can apply smoothing splines or fit a local polynomial regression to the observed values (Y_j) and use this to estimate the noiseless trajectories (e.g., Wu et al., 2014; Henderson and Michailidis, 2014). Unfortunately, a naïve application of this strategy results in many errors. Specifically, even if a smoothed Y_j is a good estimate of X_j, the derivative of the smoothed Y_j values is usually a poor estimate of the derivative of X_j because even minor differences are exaggerated when taking the derivative. Witten noted that a clever strategy is to integrate both sides of equation 7 (Dattner and Klaassen, 2015), as the integral is less sensitive to small differences between Y_j and X_j and it is not difficult to estimate with observed information. This leads to an equation of the form

$$X_j(t_i) - X_j(0) \approx t_i C_j + \sum_{k=1}^{p} \left[\int_0^{t_i} \psi(X_k(s)) ds \right] \cdot \theta_{jk}.$$ eq. 8

Replacing the derivative in equation 7 with an integral in equation 8 reduces the number of time points (N) necessary to recover the graph, in theory and in practice, Witten said. Extending Loh and Wainwright (2011), she established variable selection consistency—that the method consistently identifies the graph structure—for the standardized group lasso regression with errors in measured variables. The proposed method correctly identifies the parents of each node, which correspond to regulatory genes, with high probability and performs better than other methods in the literature.

Moving to her second example, Witten discussed graph estimation from neuron firing data collected continuously over time (e.g., Pillow et al., 2008), simplifying the example to only nine neurons although actual data sets contain measurements on many more neurons over large time periods. Each neuron can enhance, inhibit, or have no effect on the firing rate of other neurons, and a graphical model created from spike train data can intuitively present this information. Witten explained the Hawkes Process for modeling point processes (Hawkes, 1971), which defines an intensity function for neuron j (λ_j) representing the instantaneous probability of the neuron firing as follows:

$$\lambda_j(t) = \mu_j + \sum_{k=1}^{p} \sum_{i:t_{k,i} \leq t} \omega_{j,k}(t - t_{k,i}),$$ eq. 9

where μ_j is the background intensity, $\omega_{j,k}$ is a transfer function that encodes the effect of neuron k firing on the intensity function of neuron j, and $t_{k,i}$ is the time at which neuron k has spiked i times. The transfer function $\omega_{j,k}$ contains the critical information: if it is exactly zero, there is no graph edge connecting neurons j and k. As the functional form of $\omega_{j,k}$ is unknown, similar to the gene expression example, a set of basis functions ($\psi_1 \ldots \psi_M$) is used as an estimate. This leads to the following equation:

$$\lambda_j(t) \approx \mu_j + \sum_{k=1}^{p} \sum_{i:t_{k,i} \leq t} \left[\psi(t - t_{k,i}) \right]^T \beta_{jk}.$$ eq. 10

There is a similar equation for each of p neurons; thus, fitting this high-dimensional model requires estimating p^2 transfer functions. Witten again applied a group lasso penalty to induce sparsity in the vector (β_{jk}) of length M such that values are either exactly zero, corresponding to no edge, or completely nonzero, corresponding to a graph edge. This method has similar theoretical guarantees as in the gene expression example, specifically that the correct parent nodes are identified and that the estimated graphical models are selected with high probability from high-dimensional data sets.

Stepping back from these specific examples, Witten discussed model selection consistency in greater detail, saying that it amounts to a theoretical guarantee that as the number of time points grows, it is increasingly likely that the method will exactly recover the structure of the graph. Unfortunately, when the incorrect graph is estimated there is no theoretical backing that any of the edges are correct, so model selection consistency provides an all-or-nothing guarantee. Witten said it would be preferable to gain an understanding of the uncertainty associated with each identified edge—for example, a p-value or posterior distribution—as opposed to the entire graph. In the examples presented, Witten expressed that she was skeptical of the practical implications of model selection consistency results. She challenged the audience to imagine returning to their biological collaborators with a graphical model depicting thousands of edges derived from gene expression data covering tens of thousands of unique genes and saying that, with probability approaching 1, this graph is 100 percent correct. That statement would not be taken seriously by a collaborator because the underlying models used are at best crude approximations of complex biological processes. Model selection consistency provides theoretical grounding but requires assumptions that certainly do not hold in practice, said Witten, which points to a deep disconnect between theory and practice. Witten cautioned that overconfidence in the practical implications of theoretical results such as these may actually undermine statisticians' credibility in the eyes of biologists and other researchers. Thus, in the unsupervised learning context, Witten described graph estimation methods as tools for hypothesis generation and not for confirmatory analyses.

DISCUSSION OF LEARNING FROM TIME

Michael Kosorok, University of North Carolina, Chapel Hill

Michael Kosorok began by saying that he is cautiously more optimistic than Witten and that establishing consistency in the structure of the graphical model is an important first step. Preferable measures of uncertainty such as a false discovery rate or p-values may follow, but a lot of work done now does not take this first step of establishing model selection consistency. With consistency established, Kosorok described different aspects of the graphical model that inferences can be drawn on, including graph structure and edge direction, the magnitude and sign of model coefficients, or overall prediction or classification error of the model on a new or test data set. While there are several approaches in the literature, including establishing asymptotic normality (van de Geer et al., 2014) or conditioning on the estimated model structure (Lee et al., 2016), Kosorok emphasized that this is a critically important area for research.

Kosorok introduced the concept of rotational invariance as an alternative method to the lasso to induce sparsity in high-dimensional complex data sets. In a simple linear model with an error term ε

$$Y = \beta'X + \varepsilon, \qquad \text{eq. 11}$$

where the lasso assumes sparsity in β, which is reasonable when each feature has a distinct meaning such as different demographic variables. Kosorok posed this question: What if the meanings of the features are interchangeable, as is the case when a researcher is not interested in one specific gene but rather in looking for combinations or metagenes that affect the relevant outcome? In this context, Kosorok challenged the audience to consider an unknown rotation (M) for which the product $M\beta$ is sparse, asking specifically how this rotation could be estimated and what type of penalty would be necessary to make such an approach work. While this is a poorly defined problem, he said it may be helpful to bridge the divide between the model and estimation procedures that are available now and to consider how directly these methods address the scientific questions posed.

Shifting topics to modeling interactions between variables, Kosorok reminded the audience that the lasso can be constrained to enforce strong heredity—that an interaction term can be included if and only if the corresponding individual terms are included as well. This can be achieved as a convex optimization problem with a global maximum guaranteed (Radchenko and James, 2010) and generalized to the lasso (Haris et al., 2016). Kosorok wondered whether the proposed methods for gene time course and neural spike train data can be generalized to allow for non-parametric interactions—for example, by using tensor products of the bases—and suggested that this approach could be used to interrogate the additive structure of the model. Another way to think about interactions is as one part of a quadratic term (e.g., $(X_1 + X_2)^2$ contains the interaction X_1X_2), said Kosorok, who suggested including squared terms in the model to capture interactions and to allow slight departures from linearity. With this approach, it may be important to enforce strong heredity: (1) if either X_1 or X_1^2 are included, both must be included; and (2) if the interaction term X_1X_2 is included, both first-order and second-order terms must be included. A model containing all first-order terms, second-order terms, and pairwise interactions is preserved under arbitrary rotation and could be extended to allow for sparsity under unknown rotation, said Kosorok.

Feature selection for a nonparametric regression is a related but distinct problem, in that if a variable is removed from the model, all of its interaction terms are also removed. Thus variable selection is clearly different from removing coefficients in the model and is different from grouping. Kosorok said it is possible to perform feature selection in a consistent manner that improves prediction error but noted that there is relatively little research in this area. One technique called reinforcement learning trees (Zhu et al., 2015) adaptively selects features while generating random forests and can significantly reduce error and improve convergence rates compared to so-called greedy approaches that minimize the immediate mean squared error. Kosorok clarified that these results are for nonlinear models, whereas the lasso performs better than reinforcement learning trees in terms of prediction error for linear models.

Kosorok concluded by remarking that much work remains in the area of inference after regularization is used to simplify large, complex data sets. He encouraged researchers to decide carefully what elements of their model to perform inference tasks on, emphasizing that this decision must be clearly connected with the research goals. Although his presentation did not cover all of the methods of post-regularization inference, Kosorok noted that future methods should try to avoid relying on strong assumptions.

PANEL DISCUSSION

Following their presentations, Daniela Witten and Michael Kosorok participated in a panel discussion. A participant commented that the skepticism expressed by Witten regarding the results of unsupervised analyses in which the model being used is known to be an oversimplification should be extended to the context of supervised analyses as well. The participant noted that the supervising parameters used are often subjective opinions or estimates and are not necessarily reliable, so caution must be exercised even in supervised analyses. Given that suggested models are crude approximations of complex biological processes, the participant asked if models should be interpreted as a projection of reality onto the assumed model space rather than an actual model of biological reality. Witten answered that her use of basis expansions is predicated on that idea, but she expressed concern that the model misspecification may be more fundamental than not having the appropriate basis set—for example, if the model is not additive or the biological process is stochastic and cannot be modeled with additive differential equations. The old adage applies, said Witten, that there are known unknowns and unknown unknowns, and the latter is most concerning in the context of complex multivariate data sets. She added that it is not too difficult to develop a model for expression of one gene or firing of one neuron by specifying univariate distributions, but the real challenge is modeling the interactions within groups of genes or neurons as a multidimensional multivariate distribution. Regarding the first question, Witten responded that, in the supervised context, it is not a problem that a specific response is measured or provided to the analysts, although there will always be some irreducible error associated with that response that cannot be modeled. Nonetheless, the objective of a supervised analysis is to address reducible error as much as possible given a noisy response, whereas in the unsupervised context the response is totally unknown.

Another participant asked whether the structure of the graphical model was assumed to be stationary or was allowed to change over time. In a related comment, the participant drew the comparison to mixed effects modeling in more traditional longitudinal data settings, saying that independent replication across subjects allows inference in this context. He asked if independent replication—for example, across gene expression data collected from multiple subjects—could be used to support first-order inference in this context. Responding to the first question, Witten said her method for generating graphical models from time course data does require the assumption of stationarity in relationships over time, and allowing the graph to change form over time is a more challenging topic for future research. Answering the second question, Witten said that the method she introduced assumes that noise from each measurement time and gene is independent and identically distributed, which avoids the mixed effects formulation and allows the simplest model. She noted that one could argue that the noise would be correlated across time points or genes because of the measurement technologies, which are both valid points that will make modeling even more challenging. Kosorok asked if it would be helpful to consider parametric bootstrapping to obtain, shuffle, and reapply residual estimates that are assumed to be independent and identically distributed to see if stability is maintained. Witten answered that stability selection has been used in the literature. Referring back to Genevera Allen's presentation, Witten described the best available option as generating a rank list of graph edges based on different data samples and graph estimation techniques, though this approach still falls short of fully quantifying uncertainty.

Reiterating Witten's comment that graphical model estimation from large, complex multivariate data sets should be viewed as a way of generating hypotheses for future experimental investigation, one participant asked how the field could better communicate the uncertainty associated with these exploratory analyses. Kosorok answered that statisticians equivocate more than other fields, which may come across as less intellectually strong even though the equivocation is honest and grounded in available data. Ultimately it comes down to establishing trust and building partnerships with collaborators that leads to significant advances. Witten agreed, saying that while it is easy to get results that are incorrect, it is much more difficult to produce results that statisticians trust. Statistical training is powerful because it encourages a nuanced understanding of these issues, but this double-edged sword also presents a challenge for the field and in how statisticians interact with their collaborators. Given the large disconnect between statistical theory and practice, Witten commented that some of the methods published in the biology journals are much simpler than the newer methods being published in statistics journals. Following up on this comment, one participant asked if simple pairwise correlations across time and subjects are the types of methods Witten uses in biology papers. Witten responded that computing and thresholding correlation values is often all that is needed in practice: they can often answer a biologist's scientific question much more simply and with fewer assumptions than a more complicated method like the graphical lasso. Kosorok suggested that another approach is to use the estimated model to try to reproduce data sets and then to match moments, which is done in some stochastic processes that cannot be formulated as likelihood models.

An online participant asked if there are guidelines or methods to selecting tuning parameters for regularization methods, observing that the results of regularization techniques are highly sensitive to both the random sample used and the value of these parameters. Witten responded that, in the context of supervised learning, the gold standard is to evaluate test error on a validation data set. In the context of unsupervised analyses, she continued, there is no gold standard for selecting tuning parameters. Graph estimation is useful to generate a relatively simple representation of large, complex data sets, said Witten. If the tuning parameter is too small, the graph is unreadable because too many edges are included; if the tuning parameter is too large, there are no edges estimated. Finding the appropriate balance is a subjective decision made by the statistician and his/her collaborators based on the level of detail that is desired in the resulting graph, said Witten. Kosorok commented that tuning parameter selection is challenging and usually done in an ad hoc fashion. He did not know of rigorous methods to suggest as a more satisfying alternative and suggested that future theoretical studies that are appropriately skeptical could help inform approaches that are slightly more automated.

SELECTIVE INFERENCE IN LINEAR REGRESSION

Jonathan Taylor, Stanford University

Jonathan Taylor delivered a methodological presentation on selective inference, which he described as a compromise between exploratory and confirmatory analysis that allows testing a hypothesis suggested by the data. Using an illustrative example of mutation-induced human immunodeficiency virus (HIV) drug resistance, his analysis goal was to build an interpretable predictive model of resistance from a sequencing data set containing 633 distinct viruses with 91 different mutations occurring more than 10 times in the sample. Assuming sparsity in the 91 features is reasonable, and Taylor showed the results of an ordinary least squares fit to the data with only one or two features having large coefficient values. Taylor applied the square root lasso (Belloni et al., 2014; Sun and Zhang, 2012; Tian et al., 2015) in part because it does not require information about the level of noise in the data set to specify a theoretically justifiable tuning parameter. The lasso selected a subset of approximately 15 variables that represent mutation sites in the viral genotype, several of which are known in the HIV resistance literature. Fitting a regression model—for example, a parametric Gaussian model—using these parameters will yield p-values or confidence intervals for each variable; however, these convey no information about significance and cannot be used because the data were used to select the model. Taylor said a significant challenge of selective inference is that there are good methods for model selection, but they use the data and therefore require new tools for reporting and interpreting the significance (Benjamini, 2010).

One approach is to apply data splitting (Hurvich and Tsai, 1990; Wasserman and Roeder, 2009), in which a portion of the original 633 virus data set is used for variable selection and the remaining data are used for the inference task. Underlying this approach is the justification that the second portion of the data used for model fitting is independent of the first portion of the data used in model selection, explained Taylor. After splitting the data set, the square root lasso identified 11 variables; while there was some overlap from the first case described above, several differences emerged because different data were used in the two cases. Fitting a parametric Gaussian model with these variables to the unused portion of the data resulted in p-values or confidence intervals that, in principle, were justified, as the data used to select variables were independent of the data used to fit the model. Taylor cautioned that the implicit assumption in these measures of significance was that the model used for regression is a reasonable representation of the underlying process. This is in contrast to the use of data splitting in the context of cross-validation or estimating prediction error, which requires essentially no assumptions. Taylor broadly described a model as a collection of distributions that are drawn from the same space that the data came from, specification of which requires a decision on the part of the statistician. Only after the model has been specified can statistical concepts such as estimators, hypothesis testing, and confidence intervals be defined. While this is well established in classical statistical contexts, Taylor explained that in the data splitting context the goal is to use the data to choose one such model by conditioning on the first-stage data (Fithian et al., 2014; Tian and Taylor, 2015).

Taylor presented a graphical representation of the process of data splitting (Figure 5.2A): taking a random split (ω) from the original data set (X, y) results in the first-stage data (X_1, y_1), from which the square root lasso selects a subset of variables (E). In the example above, the subset of variables E corresponds to the 11 mutations that were used in a Gaussian parametric model to fit the remaining second-stage data. Yellow nodes correspond to random variables that are conditioned on, grey nodes correspond to random variables that are marginalized over, and blue nodes correspond to random variables that require specification of a model by the analyst. Taylor compared this graphical presentation of data splitting to the Box-Tukey paradigm for statistical inference, in which researchers collect exploratory data (X_e, y_e), perform exploratory analyses to identify variables of interest (E), and collect confirmatory data (X_c, y_c) on these variables for subsequent

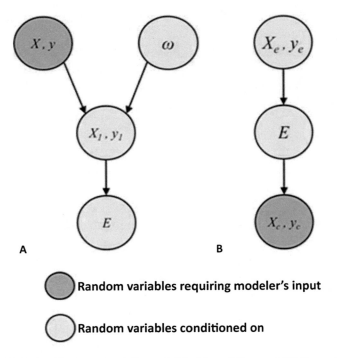

FIGURE 5.2 Graphical representations of (A) data splitting and (B) the exploratory-confirmatory analysis paradigm. X and y represent random variables before data splitting, ω represents a random split, X_1 and y_1 represent the first-stage data used to select variables E for fitting. X_e and y_e represent data collected in exploratory analysis, and X_c and y_c represent data collected in confirmatory analysis. SOURCE: Jonathan Taylor, Stanford University, "Selective inference in linear regression," presentation to the workshop, June 9, 2016.

confirmatory analysis. In the Box-Tukey paradigm, the exploratory data and variables of interest are conditioned on and model specification occurs after variables have been selected, whereas in the data splitting representation model specification occurs before variables are selected (Figure 5.2B). There is a trade-off in data splitting between the fraction of data used in the first-round screening of variables and the fraction saved for inference in the second round. Using a synthetic data set, Taylor showed that as more data are used in the first round, the statistical power of second-round inference decreases; conversely, as more data are retained for second-round inference, the probability of identifying all significant variables decreases.

STATISTICS AND BIG DATA CHALLENGES IN NEUROSCIENCE

Emery N. Brown, Massachusetts Institute of Technology,
Massachusetts General Hospital, and Harvard Medical School

Emery N. Brown began by describing his participation in the Brain Initiative, which was designed to foster development of new tools to advance the field of neuroscience. He emphasized that focusing solely on the creation of tools to generate more data was insufficient and therefore must be complemented by both statistical methods and analysis procedures to effectively utilize new data streams. Building on this motivation, Brown reviewed the Box-Tukey paradigm for statistical reasoning under uncertainty, which combines the idea of iterative model building with the delineation between exploratory and confirmatory analyses (Tukey, 1977; Box et al., 1994). Describing the Box-Tukey paradigm in the context of neuroscience (Kass et al., 2005), Brown remarked that the underlying conceptual model for neuroscience experiments is the regression model: the analyst is interested in how a subject's response to stimulus changes as different covariates are changed. In addition to being multivariate, neuroscience data are highly dynamic; however, most neuroscience data analysis methods are static, thus filtering out important information. Brown showed a few diverse examples of dynamic neuroscience data, including neuron spike train data, images produced by functional magnetic resonance imaging and diffuse optical tomography, and behavioral or cognitive performance data.

Brown presented a case study from his anesthesiology research evaluating electroencephalogram (EEG) data collected from patients under general anesthesia, which he explained is a high signal-to-noise ratio problem because background noise from movement is greatly reduced. Solving these problems may provide insights into strategies to evaluate low-signal problems that are more challenging. A video recording showed one patient's EEG signal before, during, and after administration of an anesthetic, and he called attention to the clear difference in waveforms recorded as the patient progressed from awake to anesthetized. In addition to this temporal dynamic, Brown said there are also spatial differences and patterns; he showed EEG data collected at 44 different locations on a patient's head as the dose of anesthetic is increased then decreased over a period of about 2 hours. The spectrogram at each location shows the movement of a 10 hertz (Hz) oscillation from the back of the brain to the frontal areas as anesthesia is induced, remaining in the front while the patient is unconscious and receding to the rear as drug levels decrease.

While the spatial and temporal dynamics have been known for more than 30 years, Brown and colleagues collected data on this process and combined statistical methods with experimental and modeling studies to develop a better mechanistic understanding of how the medication works. By finding the eigenvalues for the cross-spectral matrix at each frequency as a function of time and taking the ratio of the first eigenvalue to the sum of all eigenvalues, Brown calculated the global coherence of the EEG signal. If the first eigenvalue and thereby the global coherence are large, it suggests there is clear directionality to the EEG signal at that frequency and time. He showed the time course of global coherence for six patients that showed strong coherence (e.g., between 0.7 and 0.8) at 10 Hz when they were unconscious, as indicated by their lack of response to verbal questioning (Figure 5.3). Combining this observed behavior with knowledge of drug binding sites, anatomy of the brain, and modeling work suggesting that the thalamus is active over this time, Brown and colleagues inferred that the stand-

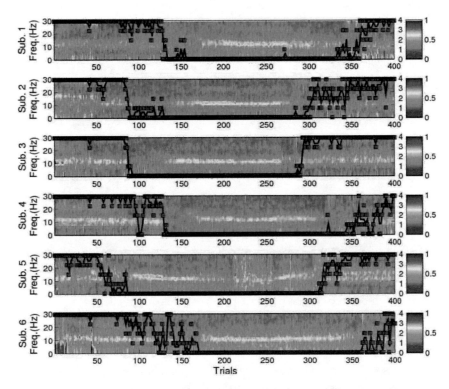

FIGURE 5.3 Time course data for six subjects shows emergence of strong coherence across the front of the scalp at approximately 10 Hz (yellow to red band, left axis) in EEG recordings while the subject is unconscious, as indicated by lack of response to yes-no questions (black line, right axis). The strong 10 Hz coherence when the subject is responding is in the posterior part of the scalp. SOURCE: Reproduced with permission from Cimenser, A., P.L. Purdon, E.T. Pierce, J.L. Walsh, A.F. Salazar-Gomez, P.G. Harrell, C. Tavares-Stoeckel, K. Habeeb, and E.N. Brown. 2011. *Proceedings of the National Academy of Sciences* 108(21): 8832-8837.

ing 10 Hz oscillation in the frontal region of the brain under anesthesia is driven by rhythmic activity between the thalamus and frontal cortex (Ching et al., 2010; Brown et al., 2011; Cimenser et al., 2011; Purdon et al., 2013). The reason this could cause unconsciousness, Brown explained, is that with slow, large amplitude oscillations, individual neurons fire infrequently, making it extremely difficult for regions of the brain to communicate. Follow-up experiments designed to test this hypothesis in rodents and monkeys verified the presence of these oscillations and phase-limited spiking activity.

Brown commented that it is often challenging to identify and apply statistical methods to answer a given question. Moving to a second illustrative example comparing the EEG spectra of cohorts of 0- to 3- and 4- to 6-month-old children under anesthesia—which appeared visually different, as the younger age group showed no oscillations above 5 Hz—he posed the seemingly simple question of whether there was a statistically significant difference in power as a function of frequency between the two age groups (Cornelissen et al., 2015). Simply plotting the power of the two groups' spectra versus frequency with 95 percent confidence intervals is sufficient for publication, but it is more appropriate to compute the difference between the two groups and construct a confidence interval around that. Brown described calculating the 95 percent confidence interval for the difference using bootstrap methods (Ramos, 1988; Hurvich and Zeger, 1987) and showed the upper and lower confidence bounds on the difference as a function of frequency. These are the curves that allow meaningful inference, said Brown, and while this is a trivial question to pose, it is not trivial to answer. Furthermore, this analysis was done for a single point in time, whereas the EEG spectra is dynamic over time; the latter presents additional challenges in identifying and applying statistically appropriate methods.

Shifting topics to statistics education, Brown said there is an unacknowledged epidemic of collective ignorance of statistics. Solutions such as delivering short courses for disciplinary scientists in statistical methods do not solve the underlying problem. It is critically important to introduce probability theory, statistics, and data analysis content earlier in education, where the curriculum accommodates repetition, so that students have sufficient time to practice and develop statistical intuition. Brown suggested that a repetitive, reinforcing curriculum be developed to introduce students in middle and high school to statistics, analogous to the way algebra is taught. Brown said this would change science dramatically. Trained statisticians are good at making inferences, and it is critical to build this capacity in students at all levels of coursework. He remarked that the fields of engineering and physics are effective at providing students with an understanding of fundamental principles, and he strives to teach concepts as fundamental principles and paradigms that are retained by students, as opposed to merely teaching a series of tests to apply based on different types of data.

Responding to a common characterization of statisticians as merely contributing to work done by their scientist collaborators, Brown asserted that statisticians are scientists too. With increasingly abundant data and the emerging field of data science, now is the time for the field of statistics to flourish. Statisticians need to consider themselves as scientists, he said, and to encourage future statisticians to become disciplinary experts or even experimentalists, pointing to the irony that statisticians are trained in design of experiments but rarely design their own experiments. Recounting personal experiences, Brown said that a high fraction of neuroscience researchers come to the field from physics and describe themselves as physicists studying neuroscience; statisticians should adopt a similar model of bringing their fundamental training to different scientific domains by becoming scientists in those domains. The field of statistics needs a credo, Brown concluded, that there is no uncertainty that statisticians cannot quantify.

DISCUSSION OF STATISTICS AND
BIG DATA CHALLENGES IN NEUROSCIENCE

Xihong Lin, Harvard University

Xihong Lin provided several examples of diverse data types that fall under the umbrella of big data, including neuroimaging, whole genome sequencing (WGS), and real-time pollution or activity monitoring collected from smart phones. She emphasized that integrating such different types of data is a critical challenge. The real value of big data is in its analysis and the inferences it allows, and the potential applications and benefits of data science are diverse. For example, in the field of health care, data science can help illuminate the causes and mechanisms of diseases, guide development of precision medicine and preventative actions, and inform health policy decisions, said Lin. She described the core of data science as being composed of statistics, computer science, and informatics, with the general goal of transforming data into knowledge (which requires inference). Reiterating that statisticians must be engaged as scientists, she pointed to the history of biostatistics as a successful engagement that resulted in better research. Lin described several lessons learned from the field of biostatistics: to always engage in cutting-edge research, to let collaboration drive method development, and to allow disciplinary scientists to be strong advocates for statisticians. She hopes that data science can continue to build on this experience, noting that it will be critical to build equally strong alliances with computer scientists and informaticians, as well as domain scientists, as each discipline brings unique and necessary skills.

Lin introduced the human genome project, launched 17 years ago with the goal of conducting WGS on the approximately 3 billion base pairs in the human genome. WGS is comprehensive compared to genome-wide association studies that cover only common variants accounting for less than 10 percent of the genome. Since the completion of the human genome project in 2003, a series of larger sequencing projects have been announced, including the 2015 National Heart, Lung, and Blood Institute's Trans-Omics for Precision Medicine Program, which will produce WGS data for 100,000 to 150,000 subjects;[2] and the 2016 National Human Genome Research Institute's Genome Sequencing Program, which plans to collect WGS data on 200,000 individuals.[3] Combined, these projects will cover over 300,000 individuals and facilitate the even more ambitious Presidential Precision Medicine Initiative, which includes WGS for 1 million people. To convey the size of the data sets resulting from such large projects, Lin explained that WGS for 300,000 people corresponds to 4×10^{16} sequenced bases—on the same order of magnitude as the number of grains of sand on a large beach. Among the 15,000 individuals who have already been sequenced, 44 percent of the single nucleotide polymorphisms (SNPs) are observed only in one person, and thus the data are characterized by high sparsity. That is, in a data matrix with WGS on 15,000 individuals (rows) and 190 million columns, 44 percent of columns have only one nonzero value.

Lin explained that the first goal of WGS is to identify genetic regions that are associated with specific diseases or traits, which is analytically challenging because of the large size of the data sets and the low signal-to-noise ratio. With so many rare variants across all people, the simplest SNP analysis is not applicable. The common rare variant analysis methods developed for dense alternatives, such as sequencing kernel estimation (Wu et al., 2011), are subject to power loss when signals are sparse in a region. In such cases, tests for sparse alternatives—for example, those based on Generalized Higher Criticism (Barnett et al., 2016; Murkerjee et al., 2015)—are necessary to increase power.

Ten years ago, Lin explained, statisticians described the "small n, large p" data set problem—in which the number of variables with available data (e.g., 20,000 gene expressions) greatly exceeds the number of observations (e.g., dozens of patients)—but now larger and larger sample sizes have created a "large n, huge p" problem. For example, the National Human Genome Research Institute's Genome Sequencing Program contains data from 200,000 subjects consisting of approximately 1 billion SNPs. Because the majority of the human genome contains rare variants, p increases with n and additional rare variants will be observed as more samples are sequenced. Thus it is critically important that statistical analysis methods are scalable, said Lin. Because each chromosome contains large intergenic regions that do not code for a specific gene (i.e., non-coding regions), it is difficult

---

[2] The website for the Trans-Omics for Precision Medicine Program is https://www.nhlbi.nih.gov/research/resources/nhlbi-precision-medicine-initiative/topmed, accessed January 6, 2017.

[3] The website for the National Human Genome Research Institute's Genome Sequencing Program is http://gsp-hg.org/, accessed January 6, 2017.

to define the appropriate unit for WGS analysis. This is one area where it will be important to incorporate known biological information, such as functional and annotation information, said Lin, which will improve the utility of results.

Moving to the importance of including statistical reasoning in the design of modern clinical trials, Lin described research using cell phone record data to construct networks representing social interactions and how this information can be used to improve HIV interventions. In such networks, random sampling may be less effective, as it is unlikely that a highly connected hub node is selected at random. A more effective strategy may be to use the data to estimate networks, to deliberately sample the hubs, and to prioritize these individuals for interventions because they come in contact with many more people. This strategy likely introduces bias that must be accounted for in analysis by developing appropriate inference procedures. It is critically important to include statisticians early in the research design process of such network-based studies, and other biological studies broadly, said Lin. This can both help address challenges such as batch effects and lead to more reproducible data generation and analysis techniques.

Lin closed by reiterating that statisticians should do the following: (1) collaborate with domain scientists on cutting-edge problems and let these interactions drive statistical method development, (2) incorporate known domain science into the inference task at hand, (3) strive to move beyond point estimation by including confidence or uncertainty statements associated with inferences from big data, and (4) make realistic assumptions and develop new methods with less stringent assumptions. Regarding the last point, Lin elaborated that many problems in biology have a low signal-to-noise ratio, but many of the methods developed for variable selection require relatively strong signals and are likely not applicable in this context. Thus more methods allowing for weak signals need to be developed. Looking toward the future, Lin said that with the increasing size of available data sets and the spread of cloud computing, it will be increasingly important to make sure that methods are scalable and computationally tractable. Related to this, Lin suggested that funding agencies consider developing infrastructure grants to help establish cloud computing resources for efficient sharing and archiving of biomedical data and beyond.

PANEL DISCUSSION

Following their presentations, Emery N. Brown and Xihong Lin participated in a panel discussion. The first participant asked the speakers how they handle the source code for any methods developed and if it is made available to the public. Brown responded that his lab makes all code freely available to the public through various repositories, although practices vary across labs, fields, and disciplines. For example, code for spectral analysis often goes to a different repository than code for neuroimaging analysis, which is shared through a consortium of repositories of which Massachusetts General Hospital is a member. Similarly, Lin said that all code from her group is made publicly available, and she reiterated that it is important for statisticians developing software to understand the practices and data formats used by domain science collaborators. Furthermore, she encouraged statisticians writing code to consider the emergence of cloud computing, as software that works well for clustered computing may not be effective in the cloud environment.

A bioinformatics master's student commented that graduate programs want students with strong computational and statistical training; however, there was little opportunity or incentive to engage with these concepts in high school or as an undergraduate biology major. Brown responded that introducing statistics earlier and more broadly in education is a critical opportunity; a fraction of the resources dedicated to training current scientists in statistics should be allocated for middle and high school students. In 5 or 10 years, these students will be postdoctoral researchers, graduate students, and undergraduates, he said, and scientific research will suffer if they are not trained properly.

6

Panel Discussion

The last session of the workshop featured a panel discussion among Alfred Hero (University of Michigan), Cosma Shalizi (Carnegie Mellon University), Andrew Nobel (University of North Carolina, Chapel Hill), Bin Yu (University of California, Berkeley) and Michael Daniels (University of Texas, Austin) that was moderated by Robert Kass (Carnegie Mellon University). The discussion reinforced many of the comments made in sessions throughout the workshop and emphasized broader concerns related to statistics and data science education, interdisciplinary collaboration, and the role of statistics in scientific discovery.

RESEARCH PRIORITIES FOR IMPROVING INFERENCES FROM BIG DATA

Kass posed a question about future priority research areas for improving inferences drawn from big data. Hero responded that, while there are many research areas where investment would advance the field, he was struck by the number of workshop presentations that focused on trying to integrate data and knowledge from different levels of description. For example, Hero described the challenge and potential value of integrating data-tracking phenomena on the subcellular and cellular levels with observational data from individual patients or cohorts. Creating models that combine these disparate types of data across different scales is a critical challenge that has many researchers stuck and does not receive sufficient attention from funding agencies. Making sound inferences from these integrative approaches will necessarily require contributions from domain scientists, statisticians, informaticians, and computer scientists, concluded Hero. Kass elaborated that this challenge is different from more classical interpretations of multiscale analyses, in which the underlying mechanisms relating phenomena are understood and relatively simple, because in this context the mechanisms relating phenomena across different scales are highly complex and potentially unknown. Nobel agreed and commented that in classical multiscale analysis there is typically a single fixed phenomena evaluated across scales, whereas the challenge identified by Hero requires evaluation of multiple phenomena across many scales. Due to the broad range of disciplinary backgrounds involved, Hero said this challenge requires the creation of large, sustained funding opportunities instead of an increase in the number of single investigator grants.

Another opportunity for funding agencies, said Yu, is to direct resources to research robustness and the implications of working with misspecified models, which is emerging in the literature but not performed in a systematic way (most studies still use one idealized model). Nobel agreed, saying that rigorous studies about model misspecification could move beyond acknowledging its existence to proving how misspecification affects downstream inference. Yu noted that simulation and computational approaches could be valuable to study dependent model structures; disciplines such as chemistry and physics have established strong computational subfields, whereas statistics has not, and she concluded that targeted investments from the National Science Foundation in computational statistics could efficiently advance understanding. Complementary to rigorous studies on model structure, Daniels said that there is a critical need to develop methods and approaches to identify and address messiness in large, heterogeneous data sets, which occurs before, and therefore underlies, model selection and inference. Although having access to additional data is a great resource, it comes with additional "messiness" with more complex causes, said Daniels, and understanding the causes and implications of this messiness will require input from multiple perspectives.

Looking toward the future, Yu imagined that there could be widespread use of artificial intelligence to automate statistical analyses for scientists who are not trained in statistics; the statistics community needs to work to ensure that appropriate methods will be incorporated into automated packages. Statistical research should be porous and outward facing, she explained, so that new ideas and challenges from domain scientists flow into the field and new statistical methods and best practices flow into the domain sciences. Yu emphasized several emerging fields of study—including causal inference and machine learning—that are frontier fields for which incorporation of statistical concepts will be critically important.

INFERENCE WITHIN COMPLEXITY AND
COMPUTATIONAL CONSTRAINTS

Moving to the next topic, Nobel introduced the general concept of "inference given complexity constraints," and he pointed to the trade-off of improving performance in the inference task at the cost of increased model, informational, and computational complexity. Regarding computational complexity, Nobel noted that inference is generally performed using a computer, and even the best-designed inference procedure is of little value if it cannot be computed. A researcher's willingness to repeat an analysis that takes 1 week is typically much less than his/her willingness to complete an analysis that takes 1 day, he continued, so efficient computing can facilitate replication and model checking. Information complexity may also present constraints, particularly given concerns regarding patient privacy that could manifest in data sets that have been randomized or have had information selectively removed. In this context, Nobel emphasized that it will be important to evaluate trade-offs between the inference task and the level of privacy protection imposed on available data. As databases continue to grow and move to cloud environments, such issues of method scalability, database management, metadata, and data sharing will become increasingly important. It is nontrivial for a group of researchers to agree on what the appropriate method and data are, let alone keep an accessible record of how each has evolved over the course of a project. While such practical considerations may not be glamorous, it is important for researchers to know and be transparent about how many permutations of data sets and methods they have tried to avoid "cherry picking" results.

EDUCATION AND CROSS-DISCIPLINARY COLLABORATION

Kass asked Yu to elaborate on the importance of cross-disciplinary collaboration in statistics. Yu said that the driving goal of statistics is to solve problems, which requires statisticians to involve domain science collaborators. She described that her research group embeds graduate students and postdoctoral scholars in domain science labs, which helps statisticians understand what questions collaborators are pursuing, how the data being evaluated are generated, and what useful knowledge is statistically supported with available data. She stated that collaborators do not always just need a p-value or confidence interval, and there is a broader opportunity to engage collaborators in creating an evolving, systematic approach to defining and pursuing statistical problems. Statisticians need to make sure that development and application of inference methods are grounded in the decision context faced by their collaborators, which may be a departure from traditional approaches. Related to this is the need for statistics students to receive communications training to improve interactions with collaborators, and Yu encouraged funding agencies to allocate resources for training in interpersonal collaboration skills within larger research grants.

Hero commented on the perception—both internal and external to the field—that statisticians are overly negative and insignificant intellectual contributors to the scientific process. He encouraged the statistics community to continue to

question the significance of findings and to provide constructive recommendations regarding potential next steps to improve confidence in research findings. In
addition to statisticians simply being more positive in interactions with collaborators, Hero suggested that targeted research investments be made in developing
statistical methods that help predict the next sequence of experiments that will
lead to improved p-values or confidence intervals. There has been some coverage
of this concept in the literature—for example, sequential design of experiments
and reinforcement learning—and these examples offer building blocks for a co-
ordinated effort, said Hero. Yu agreed, saying that statisticians must adopt a "can
do" attitude and be willing to take on hard analysis challenges without a clear idea
of how to solve them.

Shalizi commented that big data does not seem to reveal any problems with
the concept of statistical inference, but rather that big data exposes the limitations of the simplifying assumptions used in introductory statistics classes. For
example, the statistics community has always known that the linear model with
Gaussian noise is too simplified; that p-values combine information on the size of
a coefficient, how well it can be measured, and how large the sample size is but does
not indicate variable importance; and that no amount of additional data will help
if the quantity of interest is not identified in the collected variables. Nonetheless,
the community has not communicated this well outside the field and has seemingly been content to let oversimplifications from introductory statistics become
the norm. This needs to change as more researchers have access to larger data
sets—with 20 million measurements, every model coefficient that is not exactly
zero will appear to be significant. Shalizi said the statistics community needs to
think about how to convey uncertainty in these analyses and how to communicate
the meaning of parameters when a model is not correct and is misspecified. There
are ideas about how to do this within the field, but they need to be packaged so
that researchers and analysts at the lab bench or policy think tank can understand
and apply appropriate methods. If it has to be done model by model and based
on detailed mechanistic insight, it will not be scalable, said Shalizi. Yu agreed, suggesting that software programs could be automated to apply numerous tools to a
data set with very little interaction from a human.

This also suggests that statistics education has to change, not just by introducing the field to middle and high school students, but by reforming undergraduate
curricula as well, said Kass. Looking across all of the institutions teaching basic
statistics, or even limited to those taught by faculty with degrees in statistics, there
are opportunities for improvement. Researchers often approach statistics as simply trying to find the appropriate test to apply to a given data set, without deeper
consideration of underlying principles. This is in part because of how statistics is
taught, and Kass suggested that educators spend more time teaching fundamental
principles rather than a series of different tests. Yu agreed, saying that the existing

statistics curriculum is "flat" and should be reorganized in a hierarchical manner, with core principles across the curriculum leading into more in-depth topics. While many old principles still work, Yu emphasized that new ones need to be developed too. Daniels said that graduate education should provide students with experience programming and writing software. Yu agreed, saying that the best data science doctoral students should be able to program like computer science students and have formal training in both information science and communication.

Joseph Hogan (Brown University) described the necessity for the statistics curriculum to be modernized by introducing students to challenges and approaches for small n, large p data sets or for drawing causal conclusions from observational data. Some concepts may not be overly difficult to integrate into existing courses, and researchers and funding agencies need to think critically about how to improve the basic statistics curriculum. With more data science programs emerging, Hogan expressed concern that enrollment in and graduation from statistics programs could decline as the best students will be drawn to other fields. He encouraged funding agencies to develop graduate and postdoctoral training programs that specifically identify statistics as a necessary component of data science and to call out statistics explicitly in large program announcements. Shifting to future research needs, Hogan said it is critically important to make the distinction between intentionally collected data and "found data," such as electronic health records (EHRs), and he suggested that new funding opportunities be created to explore design issues that lead to meaningful inferences when using found data; this could help address challenges across many domains.

Moving to the topic of providing graduate students training in computing, Xihong Lin (Harvard University) said that many statistics students receive good training in statistical software such as R, but big data computing requires that students be exposed to additional languages and basics of software engineering, online storage and indexing platforms such as GitHub,[1] and elements of data curation and informatics. Jonathan Taylor (Stanford University) commented that good coding practices are not well rewarded in statistics departments or academia broadly and that professors need to lead by example. Lin noted that producing widely accessible statistical software often requires hiring a professional software engineer at an additional cost. In her final comment, Lin remarked on the requirement that all training grant awardees receive training in responsible conduct of research and suggested that analogous training in basic data science be considered as well. Related to this, Lin encouraged the data science community to think about what content is appropriate for a general undergraduate course for all students, similar to Harvard College's recently approved general education course called "Critical Thinking with Data."

---

[1] The website for GitHub is https://github.com/, accessed January 6, 2017.

IDENTIFICATION OF QUESTIONS AND
APPROPRIATE USES FOR AVAILABLE DATA

Recalling earlier presentations, Kass said that even when available data cannot answer a researcher's specific question, it may be possible to identify alternative questions that are well supported by available data. He encouraged further research and methods development focused on identifying such questions given a particular data set. Yu commented that there are theoretical approaches—for example, finite sample theory—for identifying what can be estimated reliably given a fixed number of observations. Another potential principle is stability, said Yu, which requires only those results that are consistent across different methods and perturbations of data to be interpreted. For example, when using clustering methods it is often unclear which method to apply, so Yu recommends applying multiple approaches and selecting only those results that are stable across all methods. Hero commented that in small n, huge p data sets, linear combinations or other patterns may be found, but parameters can be difficult to identify. He stated that methods to explore what questions can be answered are worthy of further theoretical and applied research. Daniels agreed and added that data sets with more samples than parameters—the so called large n, huge p regime—may produce deceptively small confidence intervals because the assumptions underlying the models are untested.

Genevera Allen (Rice University and Baylor College of Medicine) reminded participants of the critical difference between inference for exploratory analysis and inference for confirmatory analysis, saying that the community needs to develop new approaches and languages for communicating the high uncertainty associated with exploratory analyses. In complex data mining procedures there is high uncertainty from the data and from the methods, and statisticians need to guide domain scientists through how to interpret and use such results. Related to communication, one audience member elaborated that big data is not one homogeneous thing and that the term means different things to different people. There are easy problems to solve using big data and there are hard problems; it would be helpful to develop a taxonomy of problems big data could help solve. Relevant dimensions include the level of scientific understanding of underlying phenomena, the specific goals of the analysis, the extent of experimental control on the data used, and the ability to replicate the analysis, he said. Based on criteria such as these, the participant believes that it could be possible to identify those questions for which big data will help and those that hold little promise.

FACILITATION OF DATA SHARING AND LINKAGE

Yu urged funding agencies to help improve and incentivize data sharing—particularly referring to EHRs—across multiple institutions, saying that this remains a critical bottleneck. Data from one hospital or research center is helpful, but the real power comes from being able to combine data sets from multiple hospitals across multiple regions, she said, and Daniels agreed. Returning to the topic of model validation and broad dissemination of diagnostic tools, Yu suggested that the statistics community organize a series of discussion forums and position papers that chart a path forward and provide consensus recommendations to domain scientists working with big data regarding best statistical practices. She said other disciplines cannot be expected to avoid statistical pitfalls if the statistics community has not come to consensus on best practices. Shalizi agreed that educational activities, forums, and position papers are a good start, but these must be coupled with larger changes to the incentive structure for publishing positive findings in high-impact journals. Kass agreed, noting that this is part of a larger discussion regarding reproducible research. Hero commented that development and wide dissemination of statistics software packages could reduce the barriers to identifying and applying the appropriate tools and would advance both statistics and domain sciences.

Lin brought up the challenges of data sharing, saying that efforts need to go beyond simply sharing data by promoting linkages across different data sets. For example, it is currently difficult to link data produced by many existing large genome-wide association studies with EHR data or Medicare databases, she said, and assistance from federal agencies in achieving such data linkage could provide great resources for the research community.

THE BOUNDARY BETWEEN BIOSTATISTICS AND BIOINFORMATICS

An audience member asked the panel to elaborate on the distinction between biostatistics and bioinformatics, noting that the boundary is increasingly fluid as data management and preprocessing become more and more important to statistical analysis. Nobel responded that one distinction is that informaticians do not typically have extensive training in statistics and do not emphasize statistics in their research. Instead, informaticians typically focus on the "nuts and bolts" of working with large, high-dimensional databases, and the service they provide is essential. Yu agreed, saying a related distinction is that many bioinformatics researchers focus on solving one specific medical challenge, whereas statisticians are typically broader in their approaches. Hero, who advises students in both bioinformatics and statistics departments, observed that most bioinformatics students come from computer science and biology backgrounds, and very few have extensive math or statistics training. However, the skills and training obtained by bioinformaticians makes them essential interlocutors between statisticians and biologists or other health care professionals, concluded Hero.

References

Angrist, J.D., and W.N. Evans. 1998. Children and their parents' labor supply: Evidence from exogenous variation in family size. *The American Economic Review* 88(3): 450-477.

Barnett, I., R. Mukherjee, and X. Lin. 2016. The generalized higher criticism for testing SNP-set effects in genetic association studies. *Journal of the American Statistics Association* (accepted). http://dx.doi.org/10.1080/01621459.2016.1192039.

Bazot, C., N. Dobigeon, J.Y. Tourneret, A.K. Zaas, G.S. Ginsburg, and A.O. Hero. 2013. Unsupervised Bayesian linear unmixing of gene expression microarrays. *BMC Bioinformatics* 14. doi: 10.1186/1471-2105-14-99.

Begley, S. 2011. The best medicine. *Scientific American* 305(1): 50-55.

Belloni, A., V. Chernozhukov, and L. Wang. 2014. Pivotal estimation via square-root lasso in nonparametric regression. *Annals of Statistics* 42(2): 757-788.

Benjamini, Y. 2010. Simultaneous and selective inference: Current successes and future challenges. *Biometrical Journal* 52(6): 708-721.

Benjamini, Y., and Y. Hochberg. 1995. Controlling the false discovery rate: A practical and powerful approach to multiple testing. *Journal of the Royal Statistical Society Series B* 57(1): 289-300.

Box, G.E.P. 1979. Robustness in the strategy of scientific model building. Pp. 201-236 in *Robustness in Statistics* (R.L. Launer and G.N. Wilkinson, eds.). Academic Press Inc., New York.

Box, G.E.P., G.M. Jenkins, and G.C. Reinsel. 1994. *Time Series Analysis: Forecasting and Control.* 3rd edition. Englewood Cliffs, N.J.: Prentice Hall.

Brown, E.N., P.L. Purdon, and C.J. Van Dort. 2011. General anesthesia and altered states of arousal: A systems neuroscience analysis. *Annual Review of Neuroscience* 34: 601-628.

Bühlmann, P., and S. van de Geer. 2011. *Statistics for High-Dimensional Data: Methods, Theory, and Applications.* New York: Springer Science and Business Media.

Chen, R., G.I. Mias, J. Li-Pook-Than, L. Jiang, H.Y.K Lam, R. Chen, E. Miriami, et al. 2012. Personal omics profiling reveals dynamic molecular and medical phenotypes. *Cell* 148(6): 1293-1307.

Ching, S., A. Cimenser, P.L. Purdon, E.N. Brown, and N.J. Kopell. 2010. Thalamocortical model for a propofol-induced α-rhythm associated with loss of consciousness. *Proceedings of the National Academy of Sciences* 107(52): 22665-22670.

Cimenser, A., P.L. Purdon, E.T. Pierce, J.L. Walsh, A.F. Salazar-Gomez, P.G. Harrell, C. Tavares-Stoeckel, K. Habeeb, and E.N. Brown. 2011. Tracking brain states under general anesthesia by using global coherence analysis. *Proceedings of the National Academy of Sciences* 108(21): 8832-8837.

Cornelissen, L., S.E. Kim, P.L. Purdon, E.N. Brown, and C.B. Berde. 2015. Age-dependent electro-encephalogram (EEG) patterns during sevoflurane general anesthesia in infants. *ELife* 4. doi: 10.7554/eLife.06513.

Dattner, I., and C.A. Klaassen. 2015. Optimal rate of direct estimators in systems of ordinary differential equations linear in functions of the parameters. *Electronic Journal of Statistics* 9(2): 1939-1973.

Deng, J., W. Dong, R. Socher, L.J. Li, K. Li, and L. Fei-Fei. 2009. Imagenet: A large-scale hierarchical image database. IEEE Computer Society Conference on Computer Vision and Pattern Recognition, Miami, Fla., June 20-25.

Duchi, J.C., M.I. Jordan, and M.J. Wainwright. 2014. Privacy aware learning. *Journal of the ACM* 61(6). doi: 10.1145/2666468.

Firouzi, H., B. Rajaratnam, and A.O. Hero. 2017. Two-stage sampling, prediction and adaptive regression via correlation screening (sparcs). *IEEE Transactions on Information Theory* 63(1): 698-714.

Fithian, W., D. Sun, and J. Taylor. 2014. Optimal inference after model selection. arXiv preprint arXiv:1410.2597.

FTC (Federal Trade Commission). 2016. *Big Data: A Tool for Inclusion or Exclusion? Understanding the Issues.* https://www.ftc.gov/system/files/documents/reports/big-data-tool-inclusion-or-exclusion-understanding-issues/160106big-data-rpt.pdf.

Genberg, B.L., J.W. Hogan, and P. Braitstein. 2016. Home testing and counselling with linkage to care. *The Lancet HIV* 3(6): e244-e246.

Haneuse, S., and M. Daniels. 2016. A general framework for considering selection bias in EHR-based studies: What data are observed and why? *eGEMS (Generating Evidence & Methods to Improve Patient Outcomes)* 4(1): article 16. doi: http://dx.doi.org/10.13063/2327-9214.1203.

Haris, A., D. Witten, and N. Simon. 2016. Convex modeling of interactions with strong heredity. *Journal of Computational and Graphical Statistics* 25(4): 981-1004.

Hawkes, A.G. 1971. Spectra of some self-exciting and mutually exciting point processes. *Biometrika* 58(1): 83-90.

Henderson, J., and G. Michailidis. 2014. Network reconstruction using nonparametric additive ODE models. *PLoS ONE* 9(4). doi: 10.1371/journal.pone.0094003.

Hero, A.O., and B. Rajaratnam. 2011. Large-scale correlation screening. *Journal of the American Statistical Association* 106(496): 1540-1552.

Hero, A.O., and B. Rajaratnam. 2012. Hub discovery in partial correlation graphs. *IEEE Transactions on Information Theory* 58(9): 6064-6078.

Hero, A.O., and B. Rajaratnam. 2016. Foundational principles for large-scale inference: Illustrations through correlation mining. *Proceedings of the IEEE* 104(1): 93-110.

Hsiao, K.J., A. Kulesza, and A.O. Hero. 2014. Social collaborative retrieval. *IEEE Journal of Selected Topics in Signal Processing* 8(4): 680-689.

Huang, Y. 2011. Integrative statistical learning with applications to predicting features of diseases and health [Ph.D. thesis]. University of Michigan, Ann Arbor, Mich.

Huang, Y., A.K. Zaas, A. Rao, N. Dobigeon, P.J. Woolf, T. Veldman, N.C. Øien, et al. 2011. Temporal dynamics of host molecular responses differentiate symptomatic and asymptomatic influenza A infection. *PLoS Genetics* 7(8). doi: 10.1371/journal.pgen.1002234.

Hurvich, C.M., and C.L. Tsai. 1990. The impact of model selection on inference in linear regression. *The American Statistician* 44(3): 214-217.

Hurvich, C.M., and S. Zeger. 1987. *Frequency Domain Bootstrap Methods for Time Series.* New York: New York University.

Imai, K., G. King, and E.A. Stuart. 2008. Misunderstandings between experimentalists and observationalists about causal inference. *Journal of the Royal Statistical Society: Series A (Statistics in Society)* 171(2): 481-502.

Irizarry, R.A., C. Wang, Y. Zhou, and T.P. Speed. 2009. Gene set enrichment analysis made simple. *Statistical Methods in Medical Research* 18(6): 565-575.

Joffe, M.M., W.P. Yang, and H.I. Feldman. 2010. Selective ignorability assumptions in causal inference. *The International Journal of Biostatistics* 6(2). doi: 10.2202/1557-4679.1199.

Kass, R.E., V. Ventura, and E.N. Brown. 2005. Statistical issues in the analysis of neuronal data. *Journal of Neurophysiology* 94(1): 8-25.

Langfelder, P., P.S. Mischel, and S. Horvath. 2013. When is hub gene selection better than standard meta-analysis? *PLoS ONE* 8(4). doi: 10.1371/journal.pone.0061505.

Lee, J.D., Y. Sun, and J.E. Taylor. 2013. On model selection consistency of M-estimators with geometrically decomposable penalties. Pp. 342-350 in *Proceedings of the 26th International Conference on Neural Information Processing Systems.* Lake Tahoe, Nev., December 5-10.

Lee, J.D., D.L. Sun, Y. Sun, and J.E. Taylor. 2016. Exact post-selection inference, with application to the lasso. *Annals of Statistics* 44(3): 907-927.

Liu, T.Y., T. Burke, L.P. Park, C.W. Woods, A.K. Zaas, G.S. Ginsburg, and A.O. Hero. 2016. An individualized predictor of health and disease using paired reference and target samples. *BMC Bioinformatics* 17. doi: 10.1186/s12859-016-0889-9.

Lock, E.F., K.A. Hoadley, J.S. Marron, and A.B. Nobel. 2013. Joint and individual variation explained (JIVE) for integrated analysis of multiple data types. *Annals of Applied Statistics* 7(1): 523-542.

Lockhart, R., J. Taylor, R.J. Tibshirani, and R. Tibshirani. 2014. A significance test for the lasso. *Annals of Statistics* 42(2): 413-468.

Loh, P.L., and M.J. Wainwright. 2011. High-dimensional regression with noisy and missing data: Provable guarantees with non-convexity. Pp. 2726-2734 in *Advances in Neural Information Processing Systems* (J. Shawe-Taylor, R.S. Zemel, P.L. Bartlett, F. Pereira, and K.Q. Weinberger, eds.). Cambridge, Mass.: MIT Press.

Meng, Z., D. Wei, A. Wiesel, and A.O. Hero III. 2013. Distributed learning of Gaussian graphical models via marginal likelihoods. *Proceedings of the 16th International Conference on Artificial Intelligence and Statistics (AISTATS).* Scottsdale, Ariz., April 29-May 1. http://www.jmlr.org/proceedings/papers/v31/meng13a.pdf.

Miettinen, O.S. 1983. The need for randomization in the study of intended effects. *Statistics in Medicine* 2(2): 267-271.

Morris, A.P., B.F. Voight, T.M. Teslovich, T. Ferreira, A.V. Segre, V. Steinthorsdottir, R.J. Strawbridge, et al. 2012. Large-scale association analysis provides insights into the genetic architecture and pathophysiology of type 2 diabetes. *Nature Genetics* 44(9): 981-990.

Mukherjee, R., N.S. Pillai, and X. Lin. 2015. Hypothesis testing for high-dimensional sparse binary regression. *Annals of Statistics* 43(1): 352-381.

NITRD/NCO (National Coordination Office for Networking and Information Technology Research and Development). 2016. *The Federal Big Data Research and Development Strategic Plan.* Washington, D.C.: National Science and Technology Council. https://www.whitehouse.gov/sites/default/files/microsites/ostp/NSTC/bigdatardstrategicplan-nitrd_final-051916.pdf.

NRC (National Research Council). 2013. *Frontiers in Massive Data Analysis.* Washington, D.C.: The National Academies Press.

NRC. 2014. *Training Students to Extract Value from Big Data: Summary of a Workshop.* Washington, D.C.: The National Academies Press.

NSCI (National Strategic Computing Initiative). 2016. *National Strategic Computing Initiative Strategic Plan.* https://www.whitehouse.gov/sites/whitehouse.gov/files/images/NSCI%20Strategic%20Plan.pdf.

Pearl, J. 2000. *Causality: Models, Reasoning, and Inference.* Cambridge, U.K.: Cambridge University Press.

Pillow, J.W., J. Shlens, L. Paninski, A. Sher, A.M. Litke, E.J. Chichilnisky, and E.P. Simoncelli. 2008. Spatio-temporal correlations and visual signalling in a complete neuronal population. *Nature* 454(7207): 995-999.

Poole, D., and A.E. Raftery. 2000. Inference for deterministic simulation models: The Bayesian melding approach. *Journal of the American Statistical Association* 95(452): 1244-1255.

Purdon, P.L., E.T. Pierce, E.A. Mukamel, M.J. Prerau, J.L. Walsh, K.F.K. Wong, A.F. Salazar-Gomez, et al. 2013. Electroencephalogram signatures of loss and recovery of consciousness from propofol. *Proceedings of the National Academy of Sciences* 110(12): E1142-E1151.

Radchenko, P., and G.M. James. 2010. Variable selection using adaptive nonlinear interaction structures in high dimensions. *Journal of the American Statistical Association* 105(492): 1541-1553.

Ramos, E.A. 1988. Resampling methods for time series [PhD Dissertation]. Department of Statistics, Harvard University, Cambridge, Mass.

Rau, A., G. Marot, and F. Jaffrézic. 2014. Differential meta-analysis of RNA-seq data from multiple studies. *BMC Bioinformatics* 15. doi: 10.1186/1471-2105-15-91.

Ravikumar, P., J. Lafferty, H. Liu, and L. Wasserman. 2009. Sparse additive models. *Journal of the Royal Statistical Society: Series B (Statistical Methodology)* 71(5): 1009-1030.

Shen, R., A.B. Olshen, and M. Ladanyi. 2009. Integrative clustering of multiple genomic data types using a joint latent variable model with application to breast and lung cancer subtype analysis. *Bioinformatics* 25(22): 2906-2912.

Simon, N., and R. Tibshirani. 2012. Standardization and the group lasso penalty. *Statistica Sinica* 22(3): 983-1001.

Singh, R., J. Xu, and B. Berger. 2008. Global alignment of multiple protein interaction networks with application to functional orthology detection. *Proceedings of the National Academy of Sciences* 105(35): 12763-12768.

Smith, M. 2016. "Computer Science for All." White House Blog, January 30. https://www.whitehouse.gov/blog/2016/01/30/computer-science-all.

Song, S., K. Chaudhuri, and A.D. Sarwate. 2015. Learning from data with heterogeneous noise using SGD. *Proceedings of the 18th International Conference on Artificial Intelligence and Statistics (AISTATS).* San Diego, Calif., May 9-12. http://www.jmlr.org/proceedings/papers/v38/song15.pdf.

Sripada, C., D. Kessler, Y. Fang, R.C. Welsh, K. Prem Kumar, and M. Angstadt. 2014. Disrupted network architecture of the resting brain in attention-deficit/hyperactivity disorder. *Human Brain Mapping* 35(9): 4693-4705.

Sun, T., and C.H. Zhang. 2012. Scaled sparse linear regression. *Biometrika* 99(4). doi:10.1093/biomet/ass043.

Tian, X., and J.E. Taylor. 2015. Selective inference with a randomized response. arXiv preprint arXiv:1507.06739.

Tian, X., J.R. Loftus, and J.E. Taylor. 2015. Selective inference with unknown variance via the square-root LASSO. arXiv preprint arXiv:1504.08031.

Tukey, J.W. 1977. *Exploratory Data Analysis.* Boston, Mass.: Pearson.

van de Geer, S., P. Bühlmann, Y.A. Ritov, and R. Dezeure. 2014. On asymptotically optimal confidence regions and tests for high-dimensional models. *Annals of Statistics* 42(3): 1166-1202.

VanderWeele, T.J. 2012. Invited commentary: Structural equation models and epidemiologic analysis. *American Journal of Epidemiology* 167(7): 608-612.

Wainwright, M.J., and M.I. Jordan. 2008. Graphical models, exponential families, and variational inference. *Foundations and Trends in Machine Learning* 1(1-2): 1-305.

Wang, K., M. Narayanan, H. Zhong, M. Tompa, E.E. Schadt, and J. Zhu. 2009. Meta-analysis of inter-species liver co-expression networks elucidates traits associated with common human diseases. *PLoS Computational Biology* 5(12). doi:10.1371/journal.pcbi.1000616.

Wasserman, L., and K. Roeder. 2009. High-dimensional variable selection. *Annals of Statistics* 37(5A): 2178-2201.

Wilson, J.D., S. Wang, P.J. Mucha, S. Bhamidi, and A.B. Nobel. 2014. A testing-based extraction algorithm for identifying significant communities in networks. *Annals of Applied Statistics* 8(3): 1853-1891.

Woods, C.W., M.T. McClain, M. Chen, A.K. Zaas, B.P. Nicholson, J. Varkey, T. Veldman, et al. 2013. A host transcriptional signature for presymptomatic detection of infection in humans exposed to influenza H1N1 or H3N2. *PLoS ONE* 8(1). doi:10.1371/journal.pone.0052198.

Wu, H., T. Lu, H. Xue, and H. Liang. 2014. Sparse additive ordinary differential equations for dynamic gene regulatory network modeling. *Journal of the American Statistical Association* 109(506): 700-716.

Wu, M.C., S. Lee, T. Cai, Y. Li, M. Boehnke, and X. Lin. 2011. Rare-variant association testing for sequencing data with the sequence kernel association test. *American Journal of Human Genetics* 89(1): 82-93.

Ying, R., M. Sharma, C. Celum, J.M. Baeten, H. van Rooyen, J.P. Hughes, G. Garnett, and R.V. Barnabas. 2016. Home testing and counselling to reduce HIV incidence in a generalised epidemic setting: A mathematical modelling analysis. *The Lancet HIV* 3(6): e275-e282.

Yuan, M., and Y. Lin. 2006. Model selection and estimation in regression with grouped variables. *Journal of the Royal Statistical Society: Series B (Statistical Methodology)* 68(1): 49-67.

Zaas, A.K., T. Burke, M. Chen, M. McClain, B. Nicholson, T. Veldman, E.L Tsalik, et al. 2013. A host-based RT-PCR gene expression signature to identify acute respiratory viral infection. *Science Translational Medicine* 5(203). doi: 10.1126/scitranslmed.3006280.

Zhao, P., and B. Yu. 2006. On model selection consistency of Lasso. *Journal of Machine Learning Research* 7: 2541-2563.

Zhu, R., D. Zeng, and M.R. Kosorok. 2015. Reinforcement learning trees. *Journal of the American Statistical Association* 110(512): 1770-1784.

Zubizarreta, J.R., D.S. Small, and P.R. Rosenbaum. 2014. Isolation in the construction of natural experiments. *Annals of Applied Statistics* 8(4): 2096-2121.

Appendixes

A

Registered Workshop Participants

Abid, Rizwan – Abbas Institute of Medical Sciences, Muzaffarabad
Adams, Christopher – Federal Trade Commission
Agara Mallesh, Dhanush – University of Tennessee
Agudelo, Carlos – Ecopetrol ICP
Ahearn, Clare – Economic Consultant
Ahearn, Mary – Agricultural and Applied Economics Association
Alekseyenko, Alexander – Medical University of South Carolina
Alexiades, V. – University of Tennessee
Alharbi, Rawan – Northwestern University
Ali-Rahmani, Fatima – National Cancer Institute, National Institutes of Health (NIH)
Allen, Genevera – Rice University; Baylor College of Medicine
Allen, Melissa – Oak Ridge National Laboratory
Allen, Tim – Cigna
Alshurafa, Nabil – Northwestern University
Alves, Vinicius – University of North Carolina, Chapel Hill
Amaral, Henrique – Universidade Estadual do Maranhão
Ambani, Zainab – University of Pennsylvania
Amorim, Leila – Universidade Federal da Bahia, Brazil
Anderson, Erik – Humana
Anderson, Patrick – California Department of Public Health
Andreev, Victor – Arbor Research Collaborative for Health
Aoki, Yutaka – National Center for Health Statistics, Centers for Disease Control and Prevention (CDC)

Arterburn, David – Group Health Research Institute
Arya, Suresh – NIH
Avram, Alexandru – National Institute of Child Health and Human Development, NIH
Bagrow, James – University of Vermont
Ball, Robyn – Stanford University
Barbaresco, Frederic – Thales
Barber, Jarrett – Northern Arizona University
Barnard, Ben – Baylor University
Baro, Elande – U.S. Food and Drug Administration (FDA)
Barreto, Mauricio – Fiocruz
Baru, Chaitan – National Science Foundation
Bass, Emily – Avac
Basseville, Agnes – NIH
Basu, Saonli – University of Minnesota
Bawawana, Bavwidinsi – U.S. Census Bureau
Baydil, Banu – Columbia University
Beaudoin, Nicholas – Anser
Behan, Brendan – Ontario Brain Institute
Behera, Priyaranjan – North Carolina State University
Beresovsky, Vladislav – National Center for Health Statistics
Berlin, Brett – George Mason University
Blanken, Tessa – Netherlands Institute for Neuroscience
Boehm, Fred – University of Wisconsin, Madison
Borum, Peggy – University of Florida
Bourne, Phil – NIH
Bowles, Kathy – University of Pennsylvania School of Nursing
Bowman, Sarah – Massachusetts Institute of Technology (MIT)
Brady, Thomas – National Intrepid Center of Excellence
Bray, Mathieu – University of Michigan
Brennan, Patricia – University of Wisconsin, Madison
Brosig, Jill – Harrison Street
Brown, Emery – MIT; Massachusetts General Hospital; Harvard Medical School
Brown, Matthew – University of North Carolina, Charlotte
Bures, Regina – National Institute of Child Health and Human Development, NIH
Bushel, Pierre – National Institute of Environmental Health Sciences, NIH
Busillo, Joseph – SEI
Butaru, Florentin – Office of the Comptroller of the Currency
Cai, Wenlong – University of North Carolina, Chapel Hill
Calimlim, Brian – ICON Plc
Campbell, Robert – Brown University

Campos-Trujillo, Alfredo – Centro de Investigación en Materiales Avanzados
Cao, Nanwei – NIH
Cárdenas O'Farrill, Andrés – University of Bremen
Casper, Craig – Pikes Peak Metropolitan Planning Organization
Cassidy, Ruth – University of Michigan
Castro, David – Instituto Para la Evaluación de la Educación
Chakraborty, Hrishikesh – University of South Carolina
Chan, Melvin – National Institute of Education
Chandra, Kavitha – University of Massachusetts, Lowell
Chang, Fengshui – Old Dominion University
Chang, K.C. – George Mason University
Chang, Wo – National Institute of Standards and Technology
Chao, Ariana – University of Pennsylvania
Charlton, Sarah – U.S. Government
Chaubal, Rohan – Tata Memorial Center
Chaudhary, Mandar – North Carolina State University
Chehade, Abdallah – University of Wisconsin, Madison
Chen, Alex – Western Michigan University
Chen, Din – University of North Carolina, Chapel Hill
Chen, Gin – University of Southern California
Chen, Lily – University of Michigan
Chen, Weiping – National Institute of Diabetes and Digestive and Kidney Diseases, NIH
Chen, Weiwei – Florida International University
Chen, Yu-Chuan – National Center for Toxicological Research, FDA
Chen, Zhen – NIH
Chen, Zhuo – CDC
Cheng, Yu – University of Pittsburgh
Chernofsky, Ariel – Columbia University Mailman School of Public Health
Cheung, Paul – Xi'an Jiaotong-Liverpool University
Chiang, Chihyuan – U.S. Army Medical Research Institute of Infectious Diseases
Chih, Ming-Yuan – University of Kentucky
Cho, Eungchun – Kentucky State University
Chowbina, Sudhir – GenomeNext
Chu, Haitao – University of Minnesota
Chun, Asaph Young – U.S. Census Bureau
Chung, Matthias – Virginia Tech
Chung, Steve – California State University, Fresno
Cifuentes, Patricia – Ohio State University
Clark, William – Saint Louis University
Clegg, Lin – U.S. Department of Veterans Affairs

Coa, Kisha – ICF International
Coelho, Joseph – Marquette University
Coffman, Donna – Pennsylvania State University
Coletta, Christopher – NIH
Colopy, Glen Wright – University of Oxford
Commichaux, Seth – Georgetown University
Compher, Charlene – University of Pennsylvania
Conlin, Michael – Level X Talent
Coogan, John – Raytheon
Cooper, Gregory – University of Pittsburgh
Corbin, Amie – University of Leiden
Corliss, David – Ford Motor Company
Costa, Paulo – Instituto de Ciências Biomédicas Abel Salazar–U. Porto
Costilla, Antonio – Centro de Investigación en Matemáticas A.C.
Courtney, Paul – Dana-Farber Cancer Institute
Coyle, Patrick – NORC at the University of Chicago
Crank, Keith – National Science Foundation
Cross, Chad – Nevada State College
Cui, Naixue – University of Pennsylvania
Cui, Xinping – University of California, Riverside
Currey, Kareen – NIH
Daluwatte, Chathuri – FDA
Dani, Gabriele – Georgetown University
Daniels, Michael – University of Texas, Austin
Daramola, Olumuyiwa – Georgetown University
Dasgupta, Abhijit – National Institute of Arthritis and Musculoskeletal and Skin
    Diseases, NIH
Davey, Adam – University of Delaware
Davis, Sean – National Cancer Institute, NIH
Day, Kevin – Waggoner Engineering
De Sturler, Eric – Virginia Tech
Debakey, Samar – Health Research and Analysis
Decorte, Lauren – Accenture
Dempster, Arthur – Harvard University
Derkach, Andriy – National Cancer Institute, NIH
Deshmukh, Shraddha – University of Southern California School of Social Work
Dewey, Colin – University of Wisconsin, Madison
Deyonke, Jay – Raytheon
Dhingra, Radhika – U.S. Environmental Protection Agency (EPA)
Dibello, Anthony – National Institute of Diabetes and Digestive and Kidney
    Diseases, NIH

Dilthey, Alexander – National Human Genome Research Institute, NIH
Dixon, John – Amazon Web Services
Dolgikh, Svetlana – Kazhydromet
Dombrowski, Lad – University of Michigan
Dorfman, Alan – National Center for Health Statistics, CDC
Duncan, Dean – University of North Carolina, Chapel Hill
Dunn, Jessilyn – Stanford University
Dunn, Michelle – NIH
Dye, Laurel – National Aeronautics and Space Administration Safety Center
Eaton, Anne – Memorial Sloan Kettering Cancer Center
Eck, Michael – Army Public Health Center
Ejim, Jennifer – ITWatchIT
Ekram, Matt
Eltinge, John – U.S. Bureau of Labor Statistics
Erdley, W. Scott – Behling Simulation Center, University of Buffalo
Evain, Trevor – Lawrence Berkeley National Laboratory
Fahroo, Fariba – Defense Advanced Research Projects Agency
Fang, Jianwen – National Cancer Institute, NIH
Fann, Yang – National Institute of Neurological Disorders and Stroke, NIH
Faries, Douglas – Eli Lilly
Federer, Lisa – National Institutes of Health Library
Feng, Hao – Emory University
Fenimore, Paul – Independent Analyst
Fiaccone, Rosemeire – Federal University of Bahia
Figueroa, Patricio – World Dermatology Institute
Finneran, Kevin – National Academies of Sciences, Engineering, and Medicine
Flagan, Richard – Caltech
Flanagan, Patrick – National Resources Conservation Center, U.S. Department of
    Agriculture
Flores, Fernando – Tucma Software
Follmann, Dean – NIH
Forsythe, Alan
Fosse, Nathan – Harvard Business School
Frau, Lourdes – LMF Pharmacoepidemiology
Frey, Jeremy – University of Southampton
Fu, Yi-Ping – NIH
Gabriel-Whyte, Piriye – Canadian Executive Service Organization
Gagnon, Stuart – Federal Highway Administration
Gail, Mitchell – National Cancer Institute, NIH
Gan, Weiniu – National Heart, Lung, and Blood Institute; NIH
Gandikota, Madhuri – Tapasvi Clin-Molbio Solutions, Inc.

Gandour, Fabio – IBM Research
Garcia Islas, Luis Heriberto – Universidad Autonoma del Estado de Hidalgo
Gatsonis, Constantine – Brown University
Ge, Ping – U.S. Department of Energy
Geissert, Peter – Portland State University
George, Nysia – National Center for Toxicological Research, FDA
Gezmu, Misrak – NIH
Ghadermarzi, Shadi – Johns Hopkins University (JHU)
Ghahari, Alireza – Nuclear Energy Institute
Ghitza, Udi – National Institute on Drug Abuse, NIH
Ghosh, Basanti – Health Canada
Ghoshal, Devarshi – Lawrence Berkeley National Laboratory
Gibson, Jen – Fors Marsh Group, LLC
Gillette, Shana – U.S. Agency for International Development
Gindi, Renee – CDC
Giordano, Nicholas – University of Pennsylvania Nursing
Glaser, Elizabeth – Brandeis University
Gleicher, David – NORC at the University of Chicago
Glick, Joseph – Expertool Software, LLC
Godoy, Juan – Universidad Nacional de Córdoba
Golla, Srujana – NIH
Golozar, Asieh – Johns Hopkins Bloomberg School of Public Health
Gomes, Fabio – National Institute of Allergy and Infectious Diseases, NIH
Gordon, Chloe
Gorman, Bryan – JHU Applied Physics Laboratory
Goroff, Daniel – Alfred P. Sloan Foundation
Gouripeddi, Ram – University of Utah
Govindaiah, Swetha – University of Alabama, Huntsville
Govindarajan, Ramya – Emory University
Govindarajan, Thirupugal – NIH
Govindu, Ramakrishna – University of South Florida
Grandieri, Chris – Anne Arundel County Public Schools
Graven, Christian – North Carolina State University
Grayi, Gray
Greathouse, Leigh – Baylor University
Griffith, William – University of Washington
Groseclose, Sam – Office of Public Health Preparedness and Response, CDC
Gue, F. – Uni Konstanz
Guerra, Martin – Guerra Music Studio
Hakkinen, Pertti – National Library of Medicine, NIH
Hall, Benjamin – Humana

Hall, Henry – Consultant
Haneuse, Sebastien – Harvard University
Hanlon, Alexandra – University of Pennsylvania
Harden-Barrios, Jewel – Ochsner Health System
Harlow, Lisa – University of Rhode Island
Harris, Ann – Seton Hall University
Harris, Anna – University of Arkansas, Pine Bluff
Harris, Jason – Oak Ridge Institute for Science and Education at EPA; University
    of North Carolina
Harris, Marcus – Portland State University
Hart, B.
Hartman, Anne – Tobacco Control Research Branch, Division of Cancer Control
    and Population Sciences, National Cancer Institute, NIH
Hayes, Valerie – Hayes & Associates
He, Shisi – Georgetown University
Hector, Emily – University of Michigan
Heller, Ruth – NIH
Heller, Yair
Hero, Alfred – University of Michigan
Hewamanage, Samantha – Northwestern University
Hicks, Daniel – Science and Technology Policy Fellowships, American Association
    for the Advancement of Science
Higdon, Dave – Social and Decision Analytics Laboratory, Biocomplexity Institute
    of Virginia Tech
Higgins, Ixavier – Emory University
Hilal, Sheha – American Computer Resources
Hirschman, Karen – University of Pennsylvania School of Nursing
Hodzic, Migdat – International University of Sarajevo
Hoffeld, J. Terrell – U.S. Public Health Service
Hogan, Joseph – Brown University
Holko, Michelle – Booz Allen Hamilton
Hollm-Delgado, Maria-Graciela – JHU
Horton, Nicholas – Amherst College
Hoshizaki, Deborah – National Institute of Diabetes and Digestive and Kidney
    Diseases, NIH
Hou, Ping – University of Michigan
Howard, Rodney – National Academies of Sciences, Engineering, and Medicine
Hsu, J. – Harvard Medical School
Hsu, W. – University of California, San Diego
Hu, Ping – National Cancer Institute, NIH
Hu, Xiaowei – Oklahoma State University

Huang, Howie – George Washington University
Huang, Jie – Kaiser Permanente
Huang, Wei – Northwestern University
Huang, Weichun – National Institute of Environmental Health Sciences, NIH
Huang, Yishi – George Washington University
Hughes, Nicole – U.S. Department of Defense (DOD)
Hutfless, Susan – JHU
Huynh, Viet – Deakin University
Hyacinth, Albert – CDC
Hyun, Noorie – National Cancer Institute, NIH
Irani, Elliane – University of Pennsylvania School of Nursing
Irions, Amanda – Aim Data Science
Isayev, Olexandr – University of North Carolina, Chapel Hill
Iyer, Ganesh – Emory University
Iyer, Lax – Labcorp Inc.
Jabbar, Shirin – Emory University
Jackson, Eugenie – University of Wyoming
Jackson, Thomas – Econometrica
Jang, Don – Mathematica Policy Research
Jeng, Jessie – North Carolina State University
Ji, Chuanyi – Georgia Tech
Ji, Xiaopeng – University of Pennsylvania
Jiang, Hongmei – Northwestern University
Jiang, Miao – Harvey L. Neiman Health Policy Institute
Jiang, Zhuoxin – Medimmune
Jovanovic, Borko – Northwestern University
Juarez, Octavio – National Institute of Allergy and Infectious Diseases, NIH
Kafadar, Karen – University of Virginia
Kalmankar, Sundeep – Cisco Systems
Kang, David – Microbiotest Division of Microbac Laboratories, Inc.
Kankam, Joaquina – Prairie View A & M University Cooperative Extension Program
Kannan, Nandini – National Science Foundation
Kapphahn, Kristopher – Stanford University
Karpen, Joshua – University of California, San Diego
Kass, Robert – Carnegie Mellon University
Kaul, Abhishek – National Institute of Environmental Health Sciences, NIH
Kayaalp, Mehmet – NIH
Kelley, Douglas – GE Healthcare
Kennedy, Amy – National Cancer Institute, NIH
Khare, Meena – National Center for Health Statistics, CDC
Khatry, Deepak – Medimmune

Khattree, Ravindra – Oakland University
Khuder, Sadik – University of Toledo
Khuong, Hoa – Northeastern Illinois University
Kim, Dong-Yun – National Heart, Lung, and Blood Institute; NIH
Kim, Hyeonju – University of Arizona
Kim, Kwang-Youn – Northwestern University
Kim, Sinae – Rutgers University
Kim, Sohyoung – National Cancer Institute, NIH
Kim, Yoonsang – University of Illinois, Chicago
King, Karen – National Science Foundation
Knapp, Adam – U.S. Naval Research Laboratory
Knepper, Mark – National Heart, Lung, and Blood Institute; NIH
Knudson, Keith – Alphaport, Inc.
Ko, Yi-An – Emory University
Koch, Tara – Occupational Safety and Health Administration, U.S. Department
    of Labor
Konduri, Karthik – University of Connecticut
Kong, Lan – Pennsylvania State University, College of Medicine
Kong, Maiying – University of Louisville
Kong, Xiangrong – JHU
Kosorok, Michael – University of North Carolina, Chapel Hill
Kotsiras, Angela
Kotzinos, Dimitris – Etis Lab, University of Cergy Pontoise
Kuang, Xiaoting – Teachers College
Kuczynski, Edward – Tufts University
Kulkarni, Kshitij – University of Southern California
Kunicki, Zachary – University of Rhode Island
Kupriyanov, Roman
Kurban, Gulriz – Howard University
Kuruppumullage Don, Prabhani – Dana-Farber Cancer Institute
Labadie, Seth – U.S. Army
Laird, Douglas – U.S. Department of Transportation
Lalonde, Donna – American Statistical Association
Lancaster, Robert
Landers, Richard – Old Dominion University
Landowne, Stephen – Old Dominion University
Lanzas, Cristina – North Carolina State University
Lasater, Karen – University of Pennsylvania
Laubenbacher, Reinhard – University of Connecticut Health
Lee, George – Case Western Reserve University
Lee, Hana – Brown University

Lee, James – University of Minnesota (retired)
Lee, Julia – Northwestern University
Lee, Steven – U.S. Department of Energy Advanced Scientific Computing Research
Lee, Un Jung – National Center for Toxicological Research, FDA
Lee, W. Robert – Duke University School of Medicine
Lei, Lei
Leicht, Benjamin – U.S. Army Medical Research Institute of Infectious Diseases
Lengerich, Ben – Carnegie Mellon University
Lesko, Catherine – Johns Hopkins Bloomberg School of Public Health
Li, Alicia – Open Health Systems Laboratory
Li, Jun – University of California, Riverside
Li, Junxin – University of Pennsylvania
Li, Leping – NIH
Li, Liang – University of Texas MD Anderson Cancer Center
Li, Meng – Duke University
Li, Ming – NIH
Li, Qing – National Human Genome Research Institute, NIH
Li, Sebastian – BMO Financial Group
Li, Shijie – North Carolina State University
Li, Xiaojuan – University of North Carolina, Chapel Hill
Li, Yuanyuan – National Institute of Environmental Health Sciences, NIH
Liang, Catherine – Partners Healthcare
Liberton, Denise – NIH
Lin, Carol – CDC
Lin, Linda – Harris Corporation
Lin, Xihong – Harvard University
Lin, Yanzhu – NIH
Lin, Yong – Rutgers University
Link, Curtis – Montana Tech
Little, Roderick – University of Michigan
Liu, Anran – CDC
Liu, Bao – Fudan University
Liu, Kaibo – University of Wisconsin, Madison
Liu, Lei – Northwestern University
Liu, Rong – University of Toledo
Liu, Yusheng – Prairie View A&M University
Livermont, Elizabeth – Stevens Institute of Technology
Lobdell, Danelle – EPA
Long, Charles – Business Intelligence Services and Predictive Analytics Services
Long, Christopher – DOD
Long, Qi – Emory University

Lopez, Christy – County of San Diego Health and Human Services Agency
Lopresti, Charles – Pacific Northwest National Laboratory (retired)
Lu, Kun – University of Oklahoma
Lu, Shou-En – Rutgers University
Lu, Tsai-Ching – HRL Laboratories, LLC
Lu, Wenbin – North Carolina State University
Lubin, Jay – National Cancer Institute, NIH
Luo, Qingyang – Ochsner Health System
Luo, Yuqun – FDA
Lynch, Miranda – University of Connecticut Health Center for Quantitative Medicine
Ma, Sean – University of Michigan
Ma, Yanling – National Institute of Diabetes and Digestive and Kidney Diseases, NIH
Machado, Moara – National Cancer Institute, NIH
Madhavan, Guru – National Academies of Sciences, Engineering, and Medicine
Makambi, Kepher – Georgetown University
Makowsky, Robert – HTG Molecular
Malone, Susan – University of Pennsylvania
Manz, Boryana – PCCI
Marchant, Roman – University of Sydney
Marcotte, John – University of Michigan
Mariotto, Angela – National Cancer Institute, NIH
Markatou, Marianthi – University at Buffalo
Markowitz, David – Intelligence Advanced Research Projects Activity
Marrakchi Ben Jaafar, Ouwais – Lawrence Berkeley National Laboratory
Martinez Fonte, Leyden – Scotiabank
Martinez, Wendy – U.S. Bureau of Labor Statistics
Massie, Tammy – NIH
Matusko, Niki – University of Michigan Health System
Maziarz, Marlena – National Cancer Institute, NIH
McBee, David J. – University of Arizona
McCalla, Clement
McElroy, Charles – Case Western Reserve University
McGuire, Mary – University of Texas Medical School at Houston
McKaig, Rosemary – Division of AIDS, National Institute of Allergy and
    Infectious Diseases, NIH
McKay, Cameron – Georgetown University
McLean, Tameika – Adventist Healthcare
McManus, Doug – Freddie Mac
McNeil, Becky – U.S. Department of Veterans Affairs
Mejia, Raymond – Laboratory of Cardiac Energetics; National Heart, Lung, and
    Blood Institute; NIH

Mendes, Pedro – University of Connecticut Health
Merchant, Anand – Leidos
Messner, Michael – EPA
Meyer, Denny – Swinburne University of Technology
Meyer, Eugene – Loyola University, Maryland (retired)
Mietchen, Daniel – NIH
Miller, Peter – U.S. Census Bureau
Miller, Suzanne – Cobalt Spin
Mitchell, Kimberley
Mkondiwa, Maxwell – University of Minnesota
Mondrzejewski, Dawid – Stena
Montalvo-Urquizo, Jonathan – Center for Research in Mathematics
Moore, Taplin – U.S. Army Medical Research Institute of Infectious Diseases
Morland, Andrew – University of Wisconsin, Madison
Morris, Jeff – University of Texas MD Anderson Cancer Center
Moser, Richard – National Cancer Institute, NIH
Mughal, Zaki – University of Houston
Mukherjee, Kumar – Chicago State University
Murrie, Bruce – U.S. Department of Education (retired)
Nazimuddin Fazal, Sinan – Al-Hussan International School, Khobar, Dammam Ksa
Nemeth, Margaret – Statistical Consultants Plus, LLC
Newton, Elizabeth – Fontbonne University
Ng, Nawi – Umeå University, Sweden
Nguyen, Quynh – University of Utah
Ni, Andy – Memorial Sloan Kettering Cancer Center
Nielson, Jessica – University of California, San Francisco
Nobel, Andrew – University of North Carolina, Chapel Hill
Norman, John – ExxonMobil Biomedical Sciences, Inc.
Nowell, Lucy – U.S. Department of Energy Office of Science
Nsoesie, Elaine – University of Washington
Nussbaum, Amy – American Statistical Association
Nyamekye, Kofi – Integrated Activity-Based Simulation Research, Inc.
Ocelewski, Eric – piandpower.com
Ogawa, V. Ayano – National Academies of Sciences, Engineering, and Medicine
Oliva, Nancy – University of California, San Francisco
Oliver, Karen – Veterans Health Administration
Olson, William
Ong, Mei-Sing – Boston Children's Hospital
Oreskovich, Joanne – Montana Department of Public Health and Human Services
Ortega-Villa, Ana Maria – National Institute of Child Health and Human
    Development, NIH

Ostrouchov, George – Oak Ridge National Laboratory
O'Sullivan, Brendan – DOD
Overby, Casey – JHU
Padilla, Luis – Arizona State University
Pai, Vinay – National Institute of Biomedical Imaging and Bioengineering, NIH
Pal Choudhury, Parichoy – JHU
Palencia, Francisco – Universidad Nacional de Colombia
Pan, Wei – Duke University
Panagiotou, Orestis – National Cancer Institute, NIH
Panchal, Rekha – U.S. Army Medical Research Institute of Infectious Diseases
Pandey, Abhishek – Office of Biostatistics and Epidemiology, Center for Biologics
    Evaluation and Research, FDA
Pantula, Sastry – Oregon State University
Pao, Gerald – Salk Institute for Biological Studies
Park, Soojin – Columbia University
Patel, Pragneshkumar – University of Tennessee
Patsopoulos, Nikolaos – Harvard Medical School
Pearson, John – Duke University
Peddada, Shyamal – National Institute of Environmental Health Sciences, NIH
Pescatore, John
Pfeiffer, Ruth – National Cancer Institute, NIH
Pierson, Steve – American Statistical Association
Pino, Robinson – U.S. Department of Energy
Pita, Jorge
Plata Stapper, Andres – Stanford University
Popova, Olga – Energy Information Administration, U.S. Department of Energy
Potluru, Vamsi – Comcast Research
Pratap, Abhishek – Sage Bionetworks; University of Washington
Presnell, Brett – University of Florida
Prost-Domasky, Scott – Apes, Inc.
Pugach, Oksana – University of Illinois, Chicago
Qian, Jing – University of Massachusetts, Amherst
Qin, Steve – Emory University
Quach, Kathleen – Salk Institute
Quiroz, Antonio – Universidad Autónoma del Carmen
Racuya-Robbins, Ann – World Knowledge Bank
Radmacher, Michael – Humana
Raghavan, Vasanthan – Qualcomm
Raghavan, Vijay – Harvard University
Raghuram, Viswanathan – National Heart, Lung, and Blood Institute; NIH
Ragland, John – University of Rhode Island

Ramachandran, Mahesh – Cape Cod Commission
Rammon, Jennifer – National Center for Health Statistics, CDC
Ramsey, Gregory – TAGR2
Rappazzo, Kristen – EPA
Raval, Devesh – Federal Trade Commission
Regan, Eileen – University of Pennsylvania School of Nursing
Reyna, Cecilia – National Scientific and Technical Research Council; University of North Carolina
Reynares, Emiliano – National Scientific and Technical Research Council
Rice, Edward – National Oceanic and Atmospheric Administration
Rice, Elise – National Cancer Institute, NIH
Richardson, Lisa – CDC
Richmond, Therese – University of Pennsylvania
Rigdon, Joseph – Stanford University
Ritz, Derek – ecGroup Inc.
Riyazuddin, Firas – NIH
Rodriguez, Pedro – Facultad de Ciencias Exactas y Naturales, Universidad de Buenos Aires
Rojas García, Juan – University of Granada
Rokke, Laurie – National Oceanic and Atmospheric Administration
Rolka, Deborah – Centers for Disease Control and Prevention
Rosa, Pedro – Innopolis
Rosarda, Jessica – NIH
Rose, Jhona – Health Canada
Rose, Roderick – University of North Carolina, Chapel Hill
Rose, Sophia Miryam – Stanford University; Veterans Affairs Palo Alto Health Care System
Rosemond, Erica – National Center for Advancing Translational Sciences, NIH
Roth, Holger – NIH
Roy Choudhury, Amrita – National Center for Biotechnology Information, NIH
Ruggiero, Lucia – Ghia Global Health Advisors
Ruiz-Columbie, Arquimedes – National Wind Institute, Texas Tech University
Russell, George – Fox News Channel
S., Rajani
S., Sriram – NIH
Sagan, Philip – Sagan Consulting, LLC
Salahura, Gheorghe – Office of the Comptroller of the Currency
Sales, Anne – University of Michigan
Samai, Peter – University of North Carolina Lineberger Comprehensive Cancer Center
Sampson, Joshua – National Cancer Institute, NIH

Santana, Vilma – Federal University of Bahia
Santoro, Joseph – The Atlantic and Startup Grind
Santos, Carlos Antonio De Souza Teles – Fiocruz
Satagopan, Jaya – Memorial Sloan Kettering Cancer Center
Schechtman, Edna – Ben Gurion University of the Negev
Schenck, Natalya – Office of the Comptroller of the Currency
Schultz, Henning – BASF
Schwalbe, Michelle – National Academies of Sciences, Engineering, and Medicine
Scully, Christopher – FDA
Sen, Saunak – University of Tennessee Health Science Center
Shalizi, Cosma – Carnegie Mellon University
Shankar, Venkataraman – Pennsylvania State University
Shapiro, Aaron – Health Care
Shapiro, Danielle – Health Care
Shapiro, Mary – Health Care
Shapiro, Sandra – Health Care
Sharma, Surja – University of Maryland
Shi, Chengchun – North Carolina State University
Shi, Lan – Office of the Comptroller of the Currency
Shi, Min – National Institute of Environmental Health Sciences, NIH
Shih, Nw – University of Pennsylvania
Shim, Youn – Agency for Toxic Substances and Disease Registry
Shmagin, Boris – South Dakota State University
Si, Yajuan – University of Wisconsin, Madison
Siddique, Juned – Northwestern University
Sieber, Karl – National Institute for Occupational Safety and Health, CDC
Sihm, Jeong Sep – University of North Carolina, Greensboro
Sikka, Reita – Innovatrix
Sims, Kellie – Veterans Affairs Cooperative Studies Program Epidemiology
    Center, Durham
Sinha, Shashank – University of Michigan
Sitoula, Bibas – Oklahoma University
Slud, Eric – U.S. Census Bureau
Small, Dylan – University of Pennsylvania
Smarr, Melissa – National Institute of Child Health and Human Development, NIH
Smirnova, Ekaterina – University of Wyoming
Smith, Charles Eugene – North Carolina State University
Smith, Dan – University of Rhode Island
Smith, David – EPA
Song, Changyue – University of Wisconsin, Madison
Song, Jing – Northwestern University

Song, Lina – Harvard University
Sonkin, Dmitriy – National Cancer Institute, NIH
Sood, Akshay – University of Wisconsin, Madison
Sop-Kamga, Gaelle – Georgetown University
Sorant, Alexa – National Human Genome Research Institute, NIH
Southern, Patrick – National Science Foundation
Spahr, Judy – Main Line Health
Spencer, Michael – National Science Foundation
Spratt, Heidi – University of Texas Medical Branch
Sreekumar, Vishnu – NIH
Stites, Edward – Washington University in St. Louis
Strachan, Rodney – University of Queensland
Strassle, Paula – University of North Carolina, Chapel Hill
Strawn, George – National Academies of Sciences, Engineering, and Medicine
Stuart, Elizabeth – Johns Hopkins Bloomberg School of Public Health
Sturrock, James – Federal Highway Administration
Summers, Ronald – NIH
Sun, Jiwu
Sun, Junfeng – NIH
Sun, Xuezheng – University of North Carolina, Chapel Hill
Suzuki, Rie – University of Michigan, Flint
Szalma, Sandor – Janssen
Szelepka, Sam – U.S. Census Bureau
Szewczyk, Bill – National Security Agency
Tabachnik, Eugene
Takeda, Takako – NIH
Tan, Kay See – Memorial Sloan Kettering Cancer Center
Tan, Xinyu – University of Michigan
Tanaka, Yoko – Eli Lilly
Tang, Lu – University of Michigan
Tang, Wei – NIH
Tapley, Byron – Center for Space Research, University of Texas, Austin
Taylor, Jonathan – Stanford University
Thakkar, Rahul – Insitu, Inc.
Thomas, Fridtjof – University of Tennessee Health Science Center
Thompson, Carla – University of West Florida
Thompson, William – Northwestern University
Tiwari Dikshit, Priyanka – North Carolina State University
Tognoli, Emmanuelle – Center for Complex Systems and Brain Sciences
Totah, Deema – University of Michigan
Travillian, Ravensara – Pacific Northwest College of Allied Health Sciences

Trotta, Andrés – National University of Lanús
Trout, Kimberly – University of Pennsylvania
Ui Ghiollagain, Aine – Consultant
Ulanday, Kathleene – National Cancer Institute, NIH
Umbach, David – National Institute of Environmental Health Sciences, NIH
Uno, Hajime – Dana-Farber Cancer Institute
Uzoma, Ijeoma – U.S. Army Medical Research Institute of Infectious Diseases
Valentin, Naomi – Ryone Inc.
Vandi, Henry – CDC
Vanhove, Eric – Engility Corporation
Vardhanabhuti, Saran – Harvard T.H. Chan School of Public Health
Vattikuti, Shashaank – NIH
Vedula, Swaroop – JHU
Veiras, Hernan – Mad Mobile
Venkatachalapathy, Rajesh – Portland State University
Verma, Amit – Emory University
Viana, Marcele – Capturity
Villarreal, Maria – Centro de Investigación en Matemáticas A.C.
Villavicencio, Stephan – George Washington University
Voelker, Meta – JHU Applied Physics Laboratory
Waagen, Alexander – Hughes Research Laboratories
Wactlar, Howard – Carnegie Mellon University
Wang, Mel – Greenpeace
Wang, Ming – Pennsylvania State Hershey Medical Center
Wang, San – George Washington University
Wang, Wen – University of Michigan
Wang, Yaqun – Rutgers University
Wang, Yikai – Emory University
Wang, Yuan – University of Wisconsin, Madison
Wasserman, Emily – Pennsylvania State Hershey College of Medicine Department
    of Public Health Sciences
Wasserstein, Ronald L. – American Statistical Association
Wei, Wei – University of California, San Diego
Weidman, Scott – National Academies of Sciences, Engineering, and Medicine
Welch, Lonnie – Ohio University
Wender, Ben – National Academies of Sciences, Engineering, and Medicine
Wexler, Michael
Weylandt, Michael – Rice University
White, Don – University of Toledo
Wilkins, Ken – National Institute of Diabetes and Digestive and Kidney Diseases,
    NIH

Williams, Andre – Nemours Specialty Care
Willock, Glen Roberts – Wesleyan College
Wilson, Jim – Technology Analyst
Wilson, Lauren – National Institute of Environmental Health Sciences, NIH
Witt, Michael – DNV GL
Witten, Daniela – University of Washington
Wolz, Michael – National Heart, Lung, and Blood Institute; NIH
Wong, Charlotte – Memorial Sloan Kettering Cancer Center
Wu, Jiacheng – Yale University
Wu, Jialiang Paul
Wu, Timothy – Defense Health Agency, DOD
Wu, Wenqi – Baylor University
Wu, Zhenke – Johns Hopkins Bloomberg School of Public Health
Xia, Ashley – NIH
Xian, Xiaochen – University of Wisconsin, Madison
Xu, Hong – National Institute of Environmental Health Sciences, NIH
Xu, Jianwu – NEC Laboratories America, Inc.
Xu, Yizhen – Brown University
Xue, Cao – Northwestern University
Yadegari, Ramin – University of Arizona
Yaghouby, Farid – FDA
Yang, Chengwu – Pennsylvania State College of Medicine
Yang, Jiabei – Harvard School of Public Health
Yang, Qi – Ledios Biomedical Research, Inc.
Yang, Sinji – University of Michigan
Yang, Yandan – Javelin
Ye, Wen – University of Michigan
Yeh, Chen-Min – Salk
Yoon, So Yoon – Texas A&M University
Young, Darrell – Raytheon
Yu, Bin – University of California, Berkeley
Yu, Jenny – Mycroft
Yu, Kai – National Cancer Institute, NIH
Yu, Mandi – National Cancer Institute, NIH
Yue, Mun Sang – Brown University
Yue, Yuchen – University of Maryland
Zadrozny, Sabrina – University of North Carolina, Chapel Hill
Zhai, Ruoshui – Brown University
Zhang, Henry – National Cancer Institute, NIH
Zhang, Huikun – University of Wisconsin, Madison
Zhang, Lijun – Pennsylvania State University

Zhang, Qing – CDC
Zhang, Shun – NORC at the University of Chicago
Zhang, Tianchi – ICF International
Zhang, Tingting – University of Virginia
Zhang, Xilin – University of Michigan
Zhang, Xingyou – U.S. Census Bureau
Zhang, Xinli
Zhang, Yan – Office of the Comptroller of the Currency
Zhao, Dan – University of Illinois, Chicago
Zhao, Lihui – Northwestern University
Zhou, Ling – University of Michigan
Zhou, Qin – Memorial Sloan Kettering Cancer Center
Zhuy, Yeyi – National Institute of Child Health and Human Development, NIH
Ziemba, Robert – University of South Florida
Zimmerlin, Timothy – Automation Technologies
Zou, Shasha – University of Michigan
Zuo, Yanling – Minitab

B

Workshop Agenda

8:30 a.m.   **Welcome and Overview**

Introductions from the Co-Chairs
Michael Daniels, University of Texas at Austin
Alfred Hero, University of Michigan

Perspectives from Stakeholders
Michelle Dunn, National Institutes of Health
Nandini Kannan, National Science Foundation, Division of
Mathematical Sciences

Overview of the Workshop
Michael Daniels, University of Texas at Austin

9:40   Break

10:00   **Session I - Inference About Discoveries Based on Integration of
            Diverse Data Sets**

            Presenter: Alfred Hero, University of Michigan, to speak about
                         integrating and drawing inferences from multimodal data

Discussant: Andrew Nobel, University of North Carolina at Chapel Hill

Q&A

11:45   Lunch

12:45 p.m.   **Session I, continued**

Presenter: Genevera Allen, Rice University, to speak about
             statistical methods using medical/health case studies
Discussant: Jeffrey S. Morris, MD Anderson Cancer Center

Q&A

2:10   Break

2:30   **Session II - Inference About Causal Discoveries Driven by Large
             Observational Data**

Presenter: Joseph Hogan, Brown University, to speak about causal
             inference and decision making with health record data
             using a case study on HIV in Kenya
Discussant: Elizabeth Stuart, Johns Hopkins University

Q&A

3:55   Break

4:15   **Session II, continued**

Presenter: Sebastien Haneuse, Harvard University, to discuss
             comparative effectiveness research using electronic
             health records
Discussant: Dylan Small, University of Pennsylvania

Q&A

5:40   Adjourn Day 1

JUNE 9, 2016

8:30 a.m.   **Opening Perspectives from Stakeholders**

Chaitan Baru, National Science Foundation, Computer and
Information Science and Engineering

8:40   **Session III - Inference When Regularization Is Used to Simplify
Fitting of High-Dimensional Models**

Presenter: Daniela Witten, University of Washington, to discuss
network reconstruction from high-dimensional ordinary
differential equations
Discussant: Michael Kosorok, University of North Carolina at
Chapel Hill

Q&A

10:00   Break

10:20   **Session III, continued**

Presenter: Emery Brown, Massachusetts Institute of Technology,
to speak about using different recording methods with
high-dimensional time series
Discussant: Xihong Lin, Harvard University

Q&A

Technical/Methodological Presenter
Jonathan Taylor, Stanford University

Q&A

12:30 p.m.   Lunch

1:00   **Concluding Panel Discussion**

Moderator: Robert Kass, Carnegie Mellon University

Panelists: Alfred Hero, University of Michigan
            Bin Yu, University of California, Berkeley
            Cosma Shalizi, Carnegie Mellon University
            Andrew Nobel, University of North Carolina at Chapel
            Hill

3:00   Adjourn Workshop

C

Acronyms

AMPATH	Academic Model Providing Access to Healthcare
ART	antiretroviral therapy
BD2K	Big Data to Knowledge
CATS	Committee on Applied and Theoretical Statistics
CD4	cluster of differentiation 4 (T-cells)
DNA	deoxyribonucleic acid
EEG	electroencephalogram
EHR	electronic health record
fMRI	functional magnetic resonance imaging
GTEx	Genotype-Tissue Expression
HIV	human immunodeficiency virus
Hz	hertz
MAP	Memory and Aging Project
miRNA	micro ribonucleic acid
MRI	magnetic resonance imaging

mRNA	messenger ribonucleic acid
NIH	National Institutes of Health
NSAID	nonsteroidal anti-inflammatory drug
NSF	National Science Foundation
RNA	ribonucleic acid
ROS	Religious Orders Study
SNP	single nucleotide polymorphism
TCGA	The Cancer Genome Atlas
WGS	whole genome sequencing